NEW FOREST WALKS
a seasonal wildlife guide

NEW FOREST WALKS
a seasonal wildlife guide

Andrew Walmsley

© Andrew Walmsley, 2016

All Rights Reserved. No part of this publication may be reproduced, stored in a retrieval system, or transmitted in any form or by any means – electronic, mechanical, photocopying, recording, or otherwise – without prior written permission from the publisher

Published by Sigma Leisure – an imprint of
Sigma Press, Stobart House, Pontyclerc, Penybanc Road
Ammanford, Carmarthenshire SA18 3HP

British Library Cataloguing in Publication Data

A CIP record for this book is available from the British Library

ISBN: 978-1-85058-984-6

Typesetting and Design by: Sigma Press, Ammanford, Carms

Walk route maps and photographs: © Andrew Walmsley (unless otherwise stated)

Cover photographs: above: Fallow deer; below: Beech in autumn

Printed by: Akcent Media Ltd

Disclaimer: The information in this book is given in good faith and is believed to be correct at the time of publication. Care should always be taken when walking in hill country. Where appropriate, attention has been drawn to matters of safety. The author and publisher cannot take responsibility for any accidents or injury incurred whilst following these walks. Only you can judge your own fitness, competence and experience. Do not rely solely on sketch maps for navigation: we strongly recommend the use of appropriate Ordnance Survey (or equivalent) maps.

PREFACE

Walks in the New Forest! The very thought is enough to make the mouth water with anticipation. But consider also the potential for added interest, the benefits for mind and spirit that flow from just a little knowledge and understanding of the birds, animals, plants and other inhabitants of the natural world that live in this wonderful place.

Who could fail to be thrilled, for example, by the sight of a majestic falcon dashing over a wide expanse of gorse-splattered heath or excited by a surprise encounter with a herd of deer on a quiet New Forest lawn? Or enthralled by a magnificent, multi-coloured butterfly taking nectar from bramble blossom at the edge of a peaceful woodland clearing? Well, those with senses at least a little attuned to nature can enjoy many such experiences, for in the New Forest, almost whatever the season, the weather or time of day, there is always something to fascinate, intrigue and inspire.

Some walkers, however, might appreciate a little help along the way from prompts provided to stimulate awareness of what might be seen or heard, so included here are images and information that illuminate the often secretive world of wildlife. All are directly relevant to the accompanying walks and also, when undertaken at the corresponding time of year, to those featured in this book's companion volume, *New Forest Walks – a time traveller's guide*.

Words of caution are necessary, though, to encourage realistic expectations, for not everything detailed here will be seen during every walk as the number and variety of sightings will be significantly influenced by walkers' fieldcraft and observation skills, the relative abundance of each species, vagaries of the weather and other such considerations. And, of course, chance encounters, almost by definition, depend upon generous helpings of luck. Be aware, too, that the dates suggested here for wildlife happenings are approximations only, for precise timings are determined not by the human calendar but by ever-variable weather patterns and a host of other largely unpredictable factors.

Yet whilst accepting these provisos as inevitable, walkers, particularly those who adopt the practices outlined in the *Wildlife watching tips and tricks* chapter, should regularly be rewarded with thrilling wildlife memories.

The walks

- Walk routes and associated seasons maximise wildlife watching opportunities, although all routes can be enjoyed at any time of year
- Distances vary from a relatively modest 4 kilometres (2½ miles) up to a more ambitious 13 kilometres (8 miles)
- Convenient pub stops are often available, either on or close to the walk routes
- Some of the walks are located in relatively little visited areas of the New Forest, whilst others feature infrequently used parts of already popular places
- Time estimates are included for each walk, but further, perhaps generous, allowances should be made for periods of 'standing and watching'
- All walks feature short-cuts for those who prefer gentler exercise than that associated with the full route – they are useful, too, if time constraints are a consideration and for those excessively side-tracked by the desire to experience wildlife along the way
- Optional route extensions to places of particular wildlife or other interest are included for five walks
- The routes feature few significant hills and there are virtually no stiles or gates over which to climb
- Three walks – those for March, July and October - are suitable, at least in part, for young children in sturdy buggies
- Public transport users are well-catered for by walks accessible from railway stations or by bus. All walks also pass through, or close to, public car parks
- Numerous walk start points are located around the periphery of the New Forest, thereby removing the need for motorists to travel into the often busier, more congested centre.

Spread throughout the area, these walk routes offer the prospect of year-round enjoyment, regardless of walkers' wildlife interests and fitness levels. Plentiful opportunities are also provided to create entirely new routes by combining elements of walks from this book with those contained in *New Forest Walks – a time traveller's guide*.

I hope that walkers and wildlife enthusiasts find much to appreciate.

Andrew Walmsley
Melkridge, 2015

CONTENTS

Preface	5
The New Forest: a magnificent place for wildlife	9
The New Forest through the seasons	14
Wildlife watching tips and tricks	20
Walks overview and some advice	26
Public transport	32
Some dos and don'ts	33

The walks

January – Lengthening days and the walking year begins — 36
Godshill Cricket; Hampton Ridge; Alderhill, Sloden, Amberwood and Pitts Wood Inclosures; and Cockley Hill
12 kilometres (7½ miles) with optional short-cuts

February – Welcome signs of spring — 52
Bartley Cricket, Busketts Lawn, Furzy Lawn Inclosure, Fox Hill, Rushpole Wood and Busketts Lawn Inclosure
8.5 kilometres (5¼ miles) with an optional short-cut

March – In like a lion, out like a lamb? — 68
Brockenhurst: Balmer Lawn; Standing Hat; Pignalhill, New Copse, Frame Heath, Perrywood Haseley, Parkhill and Pignal Inclosures
7.5 kilometres (4½ miles) with an optional short-cut

April – Wild flowers, butterflies and birdsong — 82
Beaulieu Road: Shatterford, Woodfidley, Rowbarrow, Pig Bush, Ferny Crofts, Yew Tree Heath and Black Down
9 kilometres (5½ miles) with optional short-cuts

May – Spring in all its glory — 97
Burley: Turf Hill, Holmsley Bog, Castleman's Corkscrew, Holmsley Inclosure and Shappen Hill
7 kilometres (4¼ miles) with optional short-cuts

June – Long summer days made for walking   112
Crockford Bridge, Beaulieu Heath, Crockford Stream and Shipton Holms
4 kilometres (2½ miles) with an optional short-cut

July – Gentle rhythms dominate the natural world   126
Ashurst: Churchplace Inclosure and Deerleap Inclosure
5.5 kilometres (3½ miles) with an optional short-cut

August – Heathlands pink with heather blossom   139
Ashley Walk, Millersford Bottom, Godshill Inclosure, Hale Purlieu,
Millersford Plantation, Deadman Bottom and Cunninger Bottom
11.5 kilometres (7¼ miles) with optional short-cuts

September – Summer slowly gives way to autumn   156
Lyndhurst: Clay Hill; Parkhill, Denny, Little Holmhill, Pondhead and Park
Ground Inclosures
8.5 kilometres (5¼ miles) with optional short-cuts

October – Fungi flourish and fallow deer rut   172
Bolderwood Deer Sanctuary, Bolderwood Arboretum, North Oakley
Inclosure, Canadian Memorial, Highland Water and Holmhill Inclosures
8 kilometres (5 miles) with optional short-cuts

November – Woodlands ablaze with colour   189
Barrow Moor, Mark Ash Wood, Wooson's Hill and Knightwood Inclosures
6 kilometres (3¾ miles) with optional short-cuts

December – In deep mid-winter   203
Fritham: Eyeworth Pond, Fritham Plain, Sloden, Woodford Bottom, High
Corner Wood, Broomy Plain and Broomy Inclosure
13 kilometres (8 miles) with optional short-cuts

Acknowledgements   219
References   220
Index   225

THE NEW FOREST
A magnificent place for wildlife

Once memorably likened to 'a scene from medieval England, a rich tapestry of ancient woodlands, bogs and windswept heathland' and more recently said to include 'the largest area of wild, or 'unsown' vegetation in lowland Britain', the land within the original New Forest perambulation was, way back in the 11th century, set aside as a Forest in the medieval sense: a Royal Forest to be used for hunting.

Subsequent Crown ownership, jealously guarded common rights, the presence of extensive areas of acidic, infertile soil and in more recent times, amenity and nature conservation interests have over the centuries conspired to keep the New Forest safe from the worst excesses of intensive agriculture and other man-made calamities, leaving relatively intact a wide variety of remarkable landscapes. Here, marvellous pasture woodlands, roamed from

Water-lilies on a pond at Acres Down

time immemorial by deer and other animals, abut broadleaved and coniferous inclosures. Extensive tracts of heathland complement adjacent grasslands, and valley mires, bogs, gravel-bottomed streams, alder carrs, drainage channels and ponds of varying size and permanence contribute wonderful wetlands.

Many of these places are largely unspoilt and important in their own right, but taken together, they create a marvellous patchwork of habitats capable of supporting tremendous wildlife diversity. Three habitat types are, however, worthy of particular mention: the ancient pasture woodlands, or ancient, unenclosed woodlands, as they are often known; the heathlands and the valley mires. Why? Because nowhere else in lowland western Europe do they occur so extensively. Then factor in the relatively mild, south of England climate and the grazing and browsing pressure exerted by deer and commoners' stock, and the result is an area quite unlike any other found in Britain.

Almost inevitably, though, despite superficial appearances to the contrary, many of these landscapes are not entirely natural. Extensive commercial forestry inclosures, latterly often wholly coniferous, were created in the 18th, 19th and 20th centuries, frequently in places where before were open heaths and ancient, unenclosed woodlands; whilst the

Ancient, unenclosed woodland in Mark Ash Wood

Gorse and heather on Yew Tree Heath

Rowbarrow Pond – part of a wider complex of wetlands

ancient woodlands that remain have often in past centuries been systematically raided for timber. Wetland management attempts have also proliferated, particularly in the 19th and 20th centuries: drains were installed, drainage channels cut and many of the streams widened, deepened and straightened in repeated efforts to reduce seasonal flooding and reclaim land that otherwise would be unsuitable for timber production and grazing.

Early 19th century oaks in Broomy Inclosure

But how times change, for wildlife-friendly broadleaved trees are now often planted in place of felled conifers, and extensive works are being undertaken – not to the satisfaction of all local people – to reverse many of the effects of past drainage and related schemes that caused excessive erosion and reduced biodiversity. Streams, for example, have been diverted back to their old courses and meanders reinstated, stream beds have been raised, the influence of drainage channels has been reduced and heather bales and other obstructions have been placed at outflows to encourage the bogs and mires to increase in extent and create the flooding that previously was so despised but now is viewed as a vital element in the production of rich grazing for commoners' stock.

More general habitat management programmes, too, continue apace. Periodic controlled heathland burning, often undertaken in late winter after the heaths have dried but before reptiles emerge and birds start to breed, helps heathers and gorse re-generate and limits encroachment by birch, Scots pine and other colonising vegetation; whilst in the absence of burning, undesirable invasive trees are sometimes individually cut, both on the heaths and on some of the old lawns. Attempts are also made to control the spread of bracken which, if left unchecked, shades out grasses and many wild flowers; areas of scrub are selectively cleared and hollies are frequently pollarded to help extend their life and increase the amount of light reaching the woodland floor.

However, efforts to find a wholly satisfactory balance between the sometimes conflicting habitat management requirements of the recreation lobby, commoners, nature conservation interests and timber producers, are fraught with difficulty. Yet despite this, and to the credit of all concerned, much wildlife thrives in these most notable of surroundings. Indeed, within the current New Forest perambulation – rather than within the larger, National

Park boundary – around ninety bird species are reasonably frequent breeders and a further twenty, or so, are regular winter visitors or passage migrants. Tiny Dartford warblers, for example, raise families on the heaths and so do stonechats, woodlarks and intriguingly idiosyncratic nightjars; whilst hen harriers and great grey shrikes are heathland visitors in autumn and winter. Curlews, lapwings, redshanks and snipe breed in the wetter areas alongside reed buntings and occasional water rails. Kingfishers, grey wagtails and mandarin ducks inhabit the streams, whilst the woodlands boast a rich variety of birds that include great spotted and lesser spotted woodpeckers, redstarts and wood warblers.

Butterflies, particularly in summer, are often widespread and numerous – thirty, or so, species can be seen, including notably strong populations of silver-studded blues, silver-washed fritillaries, white admirals and pearl-bordered fritillaries. Dragonflies and damselflies are well-represented by twenty-five, or so, species – that is around 60% of those found in Britain – including the nationally scarce, southern damselfly and small red damselfly; whilst many other insects, such as the hornet and those that depend upon access to dead and decaying timber (so widespread, particularly in the ancient, unenclosed woodlands), are also present.

All Britain's native reptiles can be found in the New Forest: adders; grass snakes; smooth snakes; common lizards; slow worms and incredibly scarce,

Ponies at Latchmore Shade

Shafts of sunlight penetrate coniferous woodland near Standing Hat

sand lizards. A range of amphibians breed in permanent and temporary pools, including common frogs and toads, and also great crested, palmate and smooth newts. Over two thousand six hundred species of fungi have been recorded including national rarities such as the nail fungus (*Poronia punctata*) and the often huge, bearded tooth fungus (*Hericium erinaceus*). And despite often heavy grazing and trampling pressure, countless wild flower species are represented that include amongst their number the rare, bog orchid and the endemic, wild gladiolus. A huge range of lichens (including internationally important populations), mosses and ferns are present and so are five species of deer; innumerable foxes, grey squirrels and badgers; a variety of shrews, mice and voles; and a considerable number of bat species.

Impressive? Yes, certainly, so-much-so that the New Forest is of international importance for many wildlife species and is designated a Site of Special Scientific Interest (SSSI), a National Nature Reserve, a Wetland of International Importance under the Ramsar Convention, a Special Protection Area under the EC Wild Birds Directive and a Proposed Special Area of Conservation under the Habitats Directive. And since 2005, the area within the New Forest perambulation has formed a substantial proportion of the more extensive New Forest National Park.

THE NEW FOREST
through the seasons

Here in Britain, every month of the year boasts its own marvellous seasonal characteristics and nowhere is this more apparent than in the New Forest where hugely varied landscapes offer unrivalled opportunities to witness seasonal garb and wildlife activity. Of course, many people have a favourite time of year, but all are unique, all have something very special to offer.

January
Although often a difficult month for wildlife, January signals the welcome start of a new year as the countryside slowly begins its long awaited return to life. The weather, inevitably, often remains wintry, but cold, clear, sunny periods are to be enjoyed and so is the progressive increase in daylight hours. And yes, early morning mists stubbornly embrace low-lying land, yet this simply enhances the atmosphere, adding subtle charm to winter's pastel shades.

But of the changing seasons there are welcome signs as the leaves of honeysuckle and lords-and-ladies make first tentative appearances, resident birds sing with increasing vigour, hazel catkins from late in the month shimmer in the breeze and frogspawn decorates pools and drainage channels, all serving as joyous reminders of the spring to come.

February
One of the calendar's most gloriously unpredictable creatures, February rarely ceases to surprise. Frost and scatterings of snow remain an ever-present threat, yet increasingly frequent, relatively warm, sunny days tell that spring is not too far away.

Wildlife presence and behaviour reflect the month's in-between-the-seasons character, and nowhere more so than in the woods, where winter bird calls continue to be heard as flock members strive to maintain contact with one another, whilst ever-increasing bouts of song herald the gradual establishment of breeding season territories, jealously guarded domains within which individual pairs will raise their broods. Indeed, some birds – common crossbills, ravens and tawny owls, for example – might even have eggs in the nest by the end of the month.

March
'In like a lion, out like a lamb' – that is March according to one old country saying, but in relatively mild, central southern England much of this notoriously changeling month's weather more frequently belongs to spring rather than winter. Thus encouraged, resident birds ensure that dawn and dusk choruses grow in intensity and variety. Newly opened pussy willow catkins, abundant blackthorn blossom and strikingly yellow gorse blooms brighten the landscape, attracting bees and other insects to feast on freshly available nectar. Wild flowers make early appearances, butterflies that have over-wintered as adults are increasingly seen, reptiles recently active after winter slumbers bask in the sunlight and bats take to the air on milder evenings. The natural world has certainly woken.

Blackthorn blossom in early spring

April
Spring has certainly arrived, but occasional cold, frosty nights continue to remind of winter. Newborn foals are increasingly present beside proud, protective mares as early growths of new grass are enthusiastically sought. Birdsong increasingly includes melodic contributions from newly arrived spring and summer visitors, whilst dragonflies, damselflies and other insects make first flights of the year.

Breeding waders – curlews, lapwings, considerably less conspicuous snipe and increasingly scarce redshanks – enliven with calls and antics favoured, primarily wetland haunts, but it is in the woods that perhaps the greatest seasonal change occurs as spring flowers make their annual dash for growth before the leaf canopy all too quickly blocks out much of the light so essential to life below.

May
Average temperatures steadily climb, increasingly frequent sunshine brightens the landscape and spring reaches its wonderful climax. During the first half of the month, the dawn chorus reaches remarkable levels of volume, whilst that at dusk offers a more subdued alternative. Deciduous trees take on pristine blankets of pale, fresh green leaves; wild flowers for a little longer

Fresh new beech leaves beside Highland Water

bloom well in the woods and others add colour to open patches of ground amongst still sombre heathers. Insect life in all its under-stated splendour increasingly abounds, reptiles may be more frequently encountered and newborn deer will very soon experience the world for the very first time.

June
Previously noisily conspicuous curlews and lapwings increasingly fall silent as territorial behaviour and displays intended to impress potential mates give way to duties more associated with caring for growing youngsters. Other birdsong, too, particularly in woodland, gradually subsides as for many species the breeding season stutters to a close.

But as if to compensate, dragonflies and damselflies on bright days take to the wing in often impressive numbers, seemingly ever-irritated heathland stonechats scold intruders from atop sprigs of gorse, silver-studded blue butterflies enjoy their brief time in the sun, heath spotted orchids bloom and so does cross-leaved heath and bell heather, both forerunners of the pinkish-purple carpets of heather blossom still to come.

July

Spring, always so eagerly awaited, has by July all too rapidly departed and in its place are the steady, unhurried rhythms of summer. Birdsong is now largely absent as many previously vociferous creatures start their annual moult, the time when old, worn feathers are discarded and brand new replacements grown. Indeed, many birds now seem to disappear altogether, but they simply recognise that moulting wing feathers make escape from predators difficult, and wisely remain concealed, out of harms way, amongst vegetation.

Summer colour near Woodfidley

Many wild flowers continue to blossom on the heaths and around the wetter places, but the month really belongs to the insect world as butterflies, dragonflies, damselflies and myriad others greedily take centre stage, delighting onlookers with impressive powers of fight, delightful colours and intricate wing and body patterns.

August

The unmistakable scent of heather blossom lingers in the heathland air, drifts on the breeze into adjoining woodlands and attracts bees and other insects keen to gorge on the nectar. Not all have to travel far, however, for hives are placed on quieter heaths by bee-keepers anxious to be rewarded with delicious honey.

Dew-laden spiders' webs shimmer in the early morning sunlight as commoners' calls carry across the open spaces, signalling the start of the annual pony drifts. Birds relax in the sunshine, butterflies seek out bramble blossom, and dragonflies and damselflies dart erratically above pools and streams. Summer for a little longer holds sway, but there is often a hint of autumn in the cool evening air.

September

Pleasant autumnal aromas pervade the woods: the mild scent of fallen, decaying leaves and that of burgeoning crops of often mysterious fungi. Blackberries continue to ripen in woodland clearings and along ride-sides, crab apples litter the ground below ancient twisted trees, and acorns and beech-mast provide in years of plenty a huge harvest that will help sustain birds and mammals through the harsher months ahead.

Dancing in the sunlight, butterflies enjoy what for many will be last waltzes, whilst dragonflies and decreasing numbers of damselflies move with typical purpose as they hunt for insect prey. Twittering flocks of swallows congregate on overhead wires, ready for long journeys south as simultaneously, the first autumn and winter visiting birds begin to arrive. As always, there is much to see and enjoy during this, the ever-popular season of *mists and mellow fruitfulness*.

October

A time of frantic activity in the natural world, October offers many impressive wildlife watching opportunities, but few that compare with the sights and sounds of rutting fallow deer – imagine for one moment the bucks' far-carrying groans and the noise of clashing antlers, and it is as if one has been magically transported to witness this spectacular annual jamboree.

Busily scurrying about the woodland floor, grey squirrels do their utmost to avoid the melee as they compete with birds, commoners' pigs and other creatures intent on gathering fallen acorns, beech-mast, sweet chestnuts and anything else considered edible.

Fungi aplenty, meanwhile, decorate the woods with fruiting bodies of bewildering shapes, shades and sizes, all matched perfectly to their own often peculiar means of reproduction. Autumn colour on the trees is also increasingly evident, but the best is yet to come: in November.

November

An absolute firecracker of a month, November offers eye popping displays of bronze, red and gold as beech and other deciduous trees dress in gaudy autumnal attire. Often occurring around mid-month, the colourful climax lasts for only a few days before being brought to a premature end by strengthening winds that dislodge torrents of multi-coloured leaves.

Winter rapidly approaches. Bats, hedgehogs and reptiles have largely retreated to the shelter of winter hibernacula, deer continue to disperse from the rutting grounds and wild flower blooms are now rarely seen, although butterflies and dragonflies may occasionally be glimpsed, particularly during bright, mild spells. Grey squirrels and woodland birds can, however, be relied upon to entertain and amuse as they gather in the remnants of the acorn and beech-mast harvests; and berry-laden holly trees attract roaming flocks of redwings and incessantly chuckling fieldfares.

Autumn gives way to winter near Matley Wood

December

Streaked by shafts of winter sunlight, mist lingers in the woods, blurring the dark, skeletal tracery of trunks, boughs and branches. On the heaths, pale, moisture-laden air clings to the hollows. Cold, clear nights bring frost that stubbornly remains in the shadows but rapidly retreats if exposed to a little warmth. Always low in the sky, the sun, however, fails to reach much of the landscape.

Deciduous trees and many plants sleep, recharging exhausted batteries, waiting patiently for the year to turn. Yet for many birds and other creatures, winter is a time of very real hardship, a time when survival is rarely assured – autumn's nut, seed and berry crops are often severely depleted, invertebrate food is frequently locked away in frosted ground and falling temperatures take their toll on tired, ill-fed bodies. The lengthening days of the new year will, however, soon offer a little much needed relief.

WILDLIFE WATCHING
tips and tricks

Is the New Forest rich in wildlife? Absolutely! Yet even amidst such abundance, walkers should not expect multiple thrilling encounters around each and every bend in the path. Indeed, on a hot summer's day or during a wind and rain-swept foray, sightings may be limited, whilst even in ideal conditions, only a small proportion of species present may be noticed by casual passers-by. However, the development of basic observation and fieldcraft skills, use of appropriate clothing and equipment, and a little thought about weather conditions and the timing of visits will help ensure many unforgettable wildlife experiences. And alright, much of what follows will perhaps appeal primarily to dedicated wildlife enthusiasts, but many elements can be embraced by those less committed without taking anything away from the simple pleasures associated with walking in the countryside.

Clothing and equipment

Clothing
Avoid clothing that will inevitably alert wildlife to your presence. Instead, wear items that do not noisily rustle with every movement of arms or legs, and are of subdued colour – there is no need to go to camouflage extremes, but garish reds, yellows, whites and the like are best avoided.

Binoculars
Binoculars are almost essential items of equipment for those who want really good views of birds or mammals, and can also assist when watching a range of other mobile wildlife species. When making a choice from the many models available, factors to consider include:
- Magnification, as indicated by the first number quoted in the equipment specification (for example, 8x40) – between 8x and 10x are most appropriate for general use
- Light-gathering capability and consequent image brightness – these are significantly influenced by the size of the lens at the front end of the binoculars, as indicated by the second number quoted in the

specification. (Ideally, dividing the second number by the first should give a result of around 5, as in 8x40 and 10x50)
- The coating used on the lenses – the better the quality of coating, the brighter the image will be
- Field of view (the extent of the scene visible when looking through the binoculars) – this is important, particularly, for example, when tracking birds in woodland, but will often be more restricted in higher magnification models
- Close focusing capability – close focusing to within, say, 2 metres (6½ feet) is helpful for obtaining wonderful views of butterflies, dragonflies, damselflies and reptiles
- Adjustable eyepiece (diopter correction) facility – this enables lenses to be separately focused for each eye, which for many people is an invaluable aid to better viewing
- Suitability for spectacle wearers – the binoculars should be able to provide a full field of view by either folding down the rubber eyecups or by turning/pushing retractable eyecup assemblies to the 'down' position
- Durability and comfort in use – consider factors such as build quality, size, weight, fit in the hand and ease of access to the focusing mechanism.

Field guides
Access to one or more field guides is often essential for successful species identification. General guides covering commonly encountered elements of all the main wildlife groups are helpful, but the acquisition of more detailed guides relating to specific areas of interest – birds, wild flowers, butterflies, dragonflies, etc – should also be considered.

Notebook and pencil, or digital or other recorder
Enthusiasts might wish to take notes for later use when consulting field guides – many people find memory notoriously unreliable when trying to recall fine details so essential to successful species identification. A notebook and pencil will suffice for this purpose, although digital and other recorders have become increasingly popular.

Weather conditions
Many birds, mammals, insects and reptiles prefer to find shelter during periods of heavy or prolonged rain, and often have little enthusiasm for activity when the wind is strong, which produces significant challenges for

those brave enough, or unfortunate enough, to be outdoors attempting to 'wildlife watch' in those conditions. (Insects and reptiles, in particular, favour warm, calm weather, although the latter also usually avoid extreme heat).

Timing of visits

Seasonal wildlife presence and behavioural traits are indicated in the individual species accounts, so there is only a need here to consider the times of day that will increase the chances of observing wildlife.

Birds

As a general rule, birds are at their most conspicuous early in the morning and, to a lesser extent, not long before darkness falls – particularly during dawn and dusk choruses, and the associated frenzies of territorial activity. There are, however, exceptions. For example, many birds during the breeding season are likely to be relatively conspicuous throughout the day; some birds of prey are often most visible from mid-morning as they soar aloft on columns of warm, rising air or, in the case of hobbies, hunt for insects; and birds during spells of harsh winter weather may of necessity search for food during almost all the hours of daylight. And, of course, crepuscular and nocturnal species, such as woodcock, nightjar and the owls, are best sought at dusk.

A buzzard soars on the thermals

Mammals

Many wild mammals found in the New Forest, with the notable exception of grey squirrels and rabbits, know instinctively that people potentially pose a threat, and have consequently adopted crepuscular and nocturnal lifestyles. However, deer and some of the other animals, as indicated in the individual species accounts, can be seen during the hours of daylight, particularly early in the morning and as dusk approaches.

Insects and reptiles
Butterflies, dragonflies, damselflies and many other insects typically take to the wing some considerable time after dawn, whilst reptiles are also most often seen after the cool, early morning air has warmed.

Silver-washed fritillary butterflies are most conspicuous on bright, sunny days

Observation skills
If wildlife along the way is to be consistently noticed, those – and this really is stating the obvious – who take the trouble to regularly look around, remain alert and keep eyes and ears tuned to the possibility of wildlife presence will enjoy the most success. An unhurried pace is also to be recommended, one that makes time available for regular stops to look and listen – speed walking and wildlife watching are not really compatible.

Fieldcraft techniques
Many wildlife species possess exceptionally keen senses of sight, hearing and, in some cases, smell, and will quickly take cover or move away from suspected sources of danger. However, a range of fieldcraft techniques are available to help reduce the risk of prompting escape.

Walk quietly
Remember that the number of successful wildlife encounters will often be inversely proportional to the amount of noise made along the way, so at the very least, converse with companions only at low volume and not at all when approaching particularly timid creatures.

And to further minimise the sound of your approach:
- Lift up your feet – even the slightest inclination to shuffle will result in undue noise
- Keep to the grassy sides of gravel tracks – gravel is the enemy of all who wish to pass unnoticed
- Walk around puddles or risk the consequences of inevitable splashing sounds
- Avoid patches of dry leaves that otherwise will rustle underfoot
- Do not tread on fallen twigs that will make a noise as they snap.

Fallow deer often retreat when disturbed

Stay downwind
When attempting to obtain closer views of mammals, try to keep the breeze in your face as you look in their direction, as otherwise your scent may be carried towards them.

Pass unseen
If straying from the walk routes, try to avoid passing over areas of open ground where obvious human shapes and movements will startle all but the most confiding of creatures. Instead, try to follow woodland edges where your presence will be obscured against a backdrop of trees.

Do not cause alarm
When a subject has been detected and closer or more prolonged views are sought:
- Avoid sudden movements, such as animatedly pointing at whatever has been seen
- Resist the inclination to hurry forward, but instead, preferably when the subject is distracted or otherwise looks away, move slowly, smoothly and carefully

- Use trees and other natural features to conceal or obscure your outline
- Be aware that many mammals and a number of other creatures, after detecting a potential source of danger, will remain motionless until they realise that they have been seen. Snatching sideways, surreptitious glances towards them in these circumstances may not be ideal, but will often result in more prolonged subject presence
- Consider the need to keep a low profile by stooping or even crawling forward
- Remember that many birds, mammals, insects and reptiles will, to some extent, tolerate human presence, but few appreciate too close an approach. If necessary, then, remain stationary if the subject appears unduly unsettled as you move in its direction, and be prepared to back away and leave it in peace
- Try to prevent your silhouette from appearing above the skyline when attempting to obtain close views of insects on the ground or on low vegetation
- Reduce the potential for causing alarm and almost certain instant flight when watching dragonflies and damselflies by simply waiting by a section of stream or beside a pond, and letting them come to you. (Some species may even be tempted to perch on a stick, specially pushed down into the mud for their benefit, a couple of feet from where you sit).

Broad-bodied chaser dragonflies will use specially placed perches

And finally

Careful preparation and the use of patiently acquired fieldcraft skills can provide many outstanding wildlife experiences, but encounters are frequently associated with at least a little luck, the good fortune to be in the right place at just the right time. And, of course, the amount of luck enjoyed will often be significantly influenced by the amount of time spent out in the countryside.

WALKS OVERVIEW
and some advice

Ratings and estimated timings
As a guide, each walk has been allocated a 'degree of difficulty' rating:

1 – Easy walking	Up to 5.5 kilometres (3½ miles), mainly on level ground
2 – Moderate walking	Between 6 and 9 kilometres (3¾ - 5 ½ miles) with, on some walks, occasional relatively steep uphill sections
3 – Quite strenuous walking	Over 6 kilometres with some quite steep uphill sections

Estimated times to allow for completion are also provided, although only for general guidance as individual walking speeds will clearly vary. Be sure, then, to use the estimates with care, particularly if you are likely to be side-tracked by wildlife encountered along the way, for birds, animals, insects and the like rarely respect human timetables – some, perhaps the minority, will immediately show themselves, whilst others may be frustratingly difficult to observe without time-consuming effort. And always allow enough time to finish the routes in daylight, for finding the way in a darkened wood, for example, can be extremely difficult.

Alternative routes
Short-cuts back to walk start points offer a range of attractive options for those who at the outset prefer to reduce the distances covered and also for those who find that wildlife watching causes unexpected delays to progress. The walks for March, April, June, July and September also include optional extensions to places likely to provide particular wildlife interest.

All walks additionally connect with (or pass close to) other routes detailed in this book and/or in *New Forest Walks - a time traveller's guide*, which

Allow plenty of time to complete the walks before darkness falls

provides opportunities to combine elements from multiple routes to create new walks tailored to individual requirements and preferences.

Children's buggies
The suitability of each route for children's buggies is indicated. For this purpose, sturdy, often 3-wheeled, off-road buggies suitable for use in the countryside are assumed, rather than the lighter-weight models designed for pavement use.

Footwear
Always remember that parts of the New Forest, particularly in winter and after rain, can be very wet and muddy, and that even the smallest of streams can temporarily become so swollen that crossing may prove difficult. It is therefore advisable, apart from when using routes that are wholly along gravel tracks, to wear strong, preferably waterproof footwear.

Take care (and an Ordnance Survey map)
Fully numbered directions and correspondingly numbered route maps are provided, but walkers are reminded that there is always a need for care and

attention to avoid missing the way – there are no signposts or public footpath signs to help, although cycle tracks are often clearly marked. (Reading ahead is also good advice, as this will help provide an overall context for the walk.)

It is also strongly recommended that access to the Ordnance Survey map of the area – Explorer OL22 – should be available on the walks, not for use in preference to the provided maps, but as a source of supplementary information. Stocked in paper form by Tourist Information Centres, bookshops and local newsagents, the Ordnance Survey map will be particularly helpful should walkers stray from the intended route and for walks labelled 'Off the beaten track' – these, at least in part, use minor paths, often through ancient, unenclosed woodlands where one leafy glade can appear much the same as the next, or across heathlands that can appear equally, bafflingly similar.

Please be aware, though, that some paths shown on the Ordnance Survey map are not always visible on the ground, and that those on the ground do not always appear on the map! (This also applies to the presence, or otherwise, of footbridges across streams and drainage channels).

Also remember that reference may be made in the route directions to aspects of the landscape – the presence of patches of gorse and areas of

One leafy glade can appear much the same as the next

woodland, for example – that might significantly change following Forestry Commission management programmes involving heathland burning, bracken clearance or tree felling; and that fallen trees, autumnal leaves and summer vegetation – particularly bracken – can obscure paths and tracks that previously were clearly visible.

Access to a compass, and mobile phone for use in emergencies might also be helpful, and so might GPS facilities, particularly if combined with a detailed mapping application. Please note, though, that signal coverage in the New Forest cannot always be guaranteed.

Footpath and car park closures

Be aware, also, that the Forestry Commission occasionally temporarily restricts public access to paths and tracks whilst tree thinning or other management work is underway. Similarly, after timber harvesting operations, routes may be closed to allow reinstatement and consolidation following repeated use by heavy vehicles.

In these circumstances, access to an Ordnance Survey map is almost essential as comprehensive diversion signs may not be present.

Additionally, primarily to reduce surface damage, a number of car parks are often closed from early November until late March, so almost all walks feature at least one alternative start point. A small number of car parks may also be closed in spring to help minimise disturbance to ground nesting birds. Details of closures are available from the Forestry Commission by telephoning 023 80283141 or by checking their website.

Enjoy the walks

Whilst walkers should to be aware of the potential difficulties associated with some New Forest walks, please remember that if the way is occasionally missed or an obstacle is encountered, a willingness to shrug the shoulders and remain relaxed will help ensure continued enjoyment.

Walk maps key

Red lines with arrows	Route
Red lines without arrows	Paths and tracks not used
Green lines with arrows	Shorter routes
Pink lines with arrows	Detours to additional points of interest (March, April, June, July and September walks)
Brown lines	'A' and other roads
Black lines	Inclosure and other boundaries, and the railway
Blue lines	Streams, rivers, drainage channels
Grey and black lines together	Linear historical features
Purple lines	Power lines
P	Parking
Green background	Woodland
Blue background	Ponds
Uncoloured background	Unless otherwise indicated, primarily heathland often interspersed with wetter ground in the 'bottoms'

Map of the area

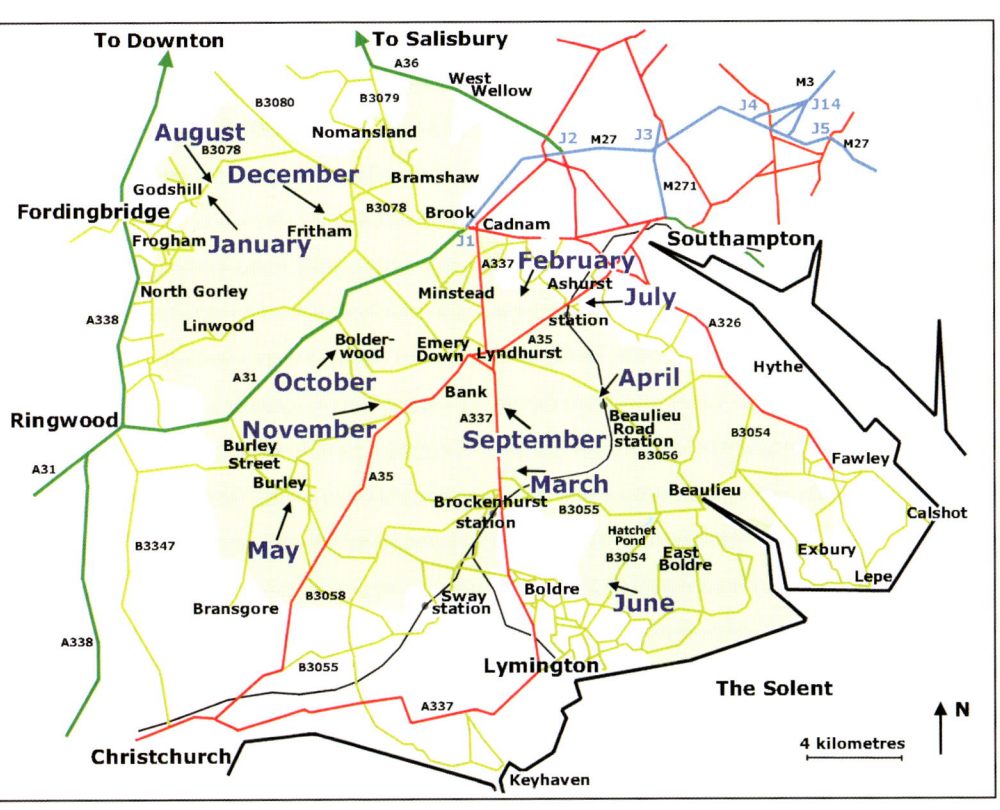

Walk months are shown in blue

Map key

Green background	The New Forest
Blue lines	Motorways
Green lines	Primary routes
Red lines	Other 'A' roads
Yellow lines	Minor roads
Black lines	Railway and coastal outline

PUBLIC TRANSPORT

Train services
The main London Waterloo to Weymouth railway line passes through the New Forest. Stations are at Ashurst (New Forest), Beaulieu Road, Brockenhurst and Sway. For details of the service, visit **www.nationalrail.co.uk**

Bus services
Bluestar buses operate from Southampton to Lymington, passing along the A35 between Ashurst and Lyndhurst; and along the A337 between Lyndhurst, Brockenhurst and Lymington. Further details are available by visiting **www.bluestarbus.co.uk**

More/Wilts and Dorset operate a number of services within the area including from Lymington to Hythe, via Beaulieu. Further details are available by visiting **www.morebus.co.uk**

National Express operates a service between London and Bournemouth. It stops once per day in each direction – at Brockenhurst and Lyndhurst. For further details check the website – **www.nationalexpress.com**

Consider, too, using a New Forest Tour Bus, a service that typically operates from late June to mid-September, currently along three circular routes:
a) The first travels from Lyndhurst to Brockenhurst and Lymington, on to Beaulieu and then Exbury and Hythe, before returning to Lyndhurst.
b) The second goes between Lyndhurst, Burley, Ringwood, Fordingbridge, Godshill, Brook, Cadnam, Woodlands and Ashurst, before returning to Lyndhurst.
c) The third route takes in sections of coastline on its way from Brockenhurst to Lymington and on to Keyhaven, Milford-on-Sea, New Milton, Bashley, Burley and back to Brockenhurst.

The Tour Buses can be flagged down along the way and will stop whenever it is safe to do so. For further information and to confirm service dates check out **www.thenewforesttour.info/**

As services are regularly reviewed and revised, visitors are advised to contact the relevant operator before travelling.

Additional information
The Public Transport Traveline can provide additional information about local public transport facilities – **www.traveline.info**

SOME DOS AND DON'TS

Apologies are probably due here to the great majority of New Forest walkers who require no reminding of the need to take care of this wonderful place, its wildlife and commoners' animals. But for the absolute avoidance of doubt:

Help safeguard New Forest wildlife
- From March to July, a range of scarce, vulnerable birds nest on or close to the ground. To avoid causing disturbance during these months, please stay on the main paths when walking on heathland or near wetlands, and if you have a dog, ensure that that does likewise
- Many other forms of wildlife are also vulnerable to changing environmental conditions, people pressure and disturbance from dogs, so please, enjoy watching wildlife, but avoid causing disturbance by trying to approach too closely – if a wild creature shows signs of agitation, immediately back away, particularly during the breeding season, a time of very real stress as young are raised
- If you find a new-born deer, leave it alone as mum will probably not be far away
- Avoid close contact with reptiles. All are harmless to people, apart, that is, from adders, that can inflict a nasty, poisonous but rarely fatal bite. Even they, though, much prefer to quietly retreat in the face of potential danger, but will respond aggressively if suddenly disturbed or if attempts are made to handle them. Good advice, then, is to leave them well alone, and reduce the likelihood of a chance encounter by staying on heathland paths where reptiles are likely to be clearly visible

Try to avoid close contact with reptiles

- Do not be tempted to pick wild flowers or uproot plants – this potentially damages the ecology of the New Forest, harms the individual species concerned, and is often illegal
- Observe the New Forest Fungi Code: there is a 1.5 kg (3 1/3 lbs) personal

collecting limit, commercial collecting is not allowed, never remove all the fungi present in an area, obey all related warning signs, and take great care with identification – some species are extremely poisonous and may even be fatal if eaten. (Some local experts believe that mushroom and toadstool populations are put at risk by even minimal picking, so even when within the 1.5 kg limit, please do not take more than is absolutely necessary for your needs).

Do not damage the archaeology or other historic features
- Do not be tempted to undertake archaeological or other digs without first consulting the New Forest National Park Authority
- Do not damage ancient New Forest earthworks or other historic features
- Never take away items of any kind from historic sites.

Take all litter home
- Litter is unsightly and, if trodden on or eaten, can have unfortunate consequences for wild animals and commoners' stock.

Remember that the commoners' animals are not pets
- Ponies and other stock animals are largely harmless, although some are not averse to nipping or kicking those who get too close. Good advice is

Commoners' animals are not pets

to always keep a sensible distance away from them. (In particular, keep well away from stallions, and also the mares when a stallion is present. Be sure, too, to never get between a stallion and his mares!)
- Do not feed the animals. It is bad for their diet, attracts them to roadsides and car parks where traffic accidents may occur, and encourages nuisance begging
- Unless instructed otherwise, close gates behind you. (To encourage the development of conditions where wild flowers and insects flourish, commoners' animals are deliberately excluded from many of the woodland inclosures. Open gates are an invitation that few animals can resist. Ponies are, however, permitted to enter a small number of inclosures where gates are locked open)
- Avoid the annual pony drifts, the late summer and autumnal animal round-ups. Casual watchers, however well-intentioned, are likely to get in the way of this serious element of commoning life
- And motorists, please be aware of the idiosyncrasies of the animals. They often frequent roadsides, but have developed little sense of the dangers associated with cars and other vehicles. Consequently, many are victims of road traffic accidents, so please be careful when driving, always keep within the speed limits, and appreciate that animal behaviour can at times be unpredictable.

Start	Godshill Cricket, Forestry Commission car park, 0.75 kilometres (½ mile) east-north-east of Godshill on the B3078 Brook to Fordingbridge road – Ordnance Survey map reference SU182151
Distance	12 kilometres (7½ miles) Shorter walk options: 1) Reduce the distance by 4 km (2½ miles) 2) Reduce the distance by 1.75 km (1 mile) 3) Reduce the distance by 2 km (1¼ miles)
Time to allow	3-7½ hours
Refreshments	The Fighting Cocks, Godshill, is close to the start of the walk
Route	Largely along readily visible tracks, although in places – for example, from the start of the walk until Hampton Ridge – a little 'off the beaten track'
Terrain	Much level ground, but also some moderate gradients. Note – parts of this walk in winter and after heavy rain can be particularly wet underfoot. (Whilst in summer, for example, the Ditchend Brook (at the end of Section 2) and the stream in Pitts Wood Inclosure (Sections 12 - 14) often run dry; after heavy or prolonged rain both can be almost impassable without Wellington boots or similar waterproof footwear. Short detours along each will, however, usually provide access to sections narrow enough to step or jump across)
Rating	3 – in places, quite strenuous walking
Buggies	Not suitable
Railway station	Ashurst (New Forest), 19.5 kilometres (12¼ miles)
Bus service	More/Wilts and Dorset serve nearby Fordingbridge
New Forest Tour Bus	Yes
Alternative starts	Ashley Walk, Forestry Commission car park at Ordnance Survey map reference SU186156
'Camping in the Forest' Caravan and Campsites	1) Longbeech, 10.5 kilometres (6½ miles) 2) Ocknell, 11.5 kilometres (7¼ miles)

JANUARY
Lengthening days and the walking year begins
(Godshill Cricket; Hampton Ridge; Alderhill, Sloden, Amberwood and Pitts Wood Inclosures; and Cockley Hill)

Winter walks can delight the senses

The walk
The heathlands, woodlands and wetlands passed through during this walk offer much of interest for walkers and wildlife enthusiasts. Located in the rugged northern section of the New Forest, the route starts at Godshill Cricket car park, situated on the B3078 Brook to Fordingbridge road, and passes

through 12 kilometres (7½ miles) of magnificently varied, undulating countryside. Optional short-cuts are available to reduce the total distance by up to 6 kilometres (3¾ miles).

Featured wildlife

Butcher's-broom – knee holm of old

Butcher's-broom, a well-distributed but rarely abundant evergreen shrub found in many of the old New Forest woodlands, resists, not wholly successfully, the depredations of deer and commoners' stock by virtue of its prickly, flattened branches. It can grow to a height of around 90 centimetres (35 inches) and sometimes occurs in clumps up to 150 centimetres (59 inches) wide. Tiny whitish-green flowers bloom from January to April, often alongside globe-shaped, bright red berries from the previous year.

Butcher's-broom often grows in large clumps

A 17th century reference suggests that the rigid stems, tied in bundles, were used to sweep butcher's blocks, a practice that gave the plant its name. It was, though, just as appropriately known as knee holm, a name that reflected its height and prickly similarity to holly – which was sometimes known as holm.

Buzzards – part of a spectacular success story

Probably the New Forest's most frequently encountered bird of prey, the large, hook-billed buzzard is a creature of fierce appearance. Often seen soaring high in the sky on broad, outstretched wings, at times

A buzzard being mobbed by a magpie

almost lost to view, or else hovering on hill-side updrafts, buzzards almost invariably attract attention with loud, far-carrying, mewing calls.

These birds have not always been commonplace, however, for numbers were relatively low until significant local and national increases occurred from the late 1980s following reduced persecution by man, the recovery of rabbit populations from the effects of myxomatosis and an end to the disastrous impact of organochlorine pesticides.

Although fond of feasting on woodpigeons, other birds, rabbits, grey squirrels and a variety of small mammals, buzzards when the ground is particularly wet also hunt on short-cropped grasslands where they stand, watch and patiently wait for indications of earthworm presence, before pouncing on what for them must surely be a fairly insubstantial meal.

Most individuals have predominantly brown plumage, but remarkable variations, ranging from very dark to very pale, cause inevitable identification confusion. Nests, predictably robust structures, are typically located high in a coniferous tree.

Fallow deer – always a joy to encounter

Of the five deer species present in the New Forest – fallow, muntjac, red, roe and sika – fallow deer are the most abundant. How many are present? Well, that is not an easy question to answer for accurate assessments are complicated by the large size of the herd, the generally reclusive nature of these creatures, their habit of regularly moving between the Crown Lands and nearby private lands, and natural peaks and troughs in population levels. However, estimates of the numbers present in spring are available, based on a visual census undertaken at that time, corrected to better reflect the likely population – the 2012 and 2013 censuses, after correction, suggested a total of at least 1,700 and 1,300 animals, respectively, whilst the Forestry Commission's strategy for 2014-2020 proposes a reduction to a corrected figure of around 950-1,250 animals.

Phoenicians, Gauls, Ancient Britons and Romans have all been credited with introducing fallow deer to Britain, but it seems likely that the population was, at most, very low until Norman imports provided impetus for significant increase.

Bucks reach a height of up to 95 centimetres (37 inches) at the shoulder; does are a little smaller. Chestnut coloured coats with pale spots are typically worn from late spring until early autumn, whilst at other times these are replaced by thicker, warmer, winter coats that are unspotted, dark grey-brown above and pale below. Throughout the year, a dark, broad line down the relatively long tail and a pale rump patch edged with black significantly aid identification. A number of coat variations do, however, occur – boldly marked menil deer, for example, stay in summer-like coat all-year-round, whilst very dark and very pale, sometimes white, animals are also seen.

Worn by yearling bucks, animals in their second year of life, the first set of antlers are simple spikes. Mature bucks, meanwhile, sport substantial palmate headgear. Antlers are cast annually, usually between mid-April and mid-May – older bucks usually cast their antlers earlier than younger animals. New growth begins almost immediately within a covering sheath of 'velvet', a protective, furry skin that bears blood vessels through which nutrients are channelled to the fairly soft, growing bone, supplementing those separately provided by an internal blood supply.

Bucks remain 'in velvet' until around mid-August, when antler growth is completed. The internal blood supply at that time ceases and for a short period the animals are considered to be 'in tatters' as the velvet decays, peels away and is removed by rubbing against overhanging branches, saplings and other vegetation, revealing clean, 'hard' antler below.

Births, usually of single fawns, mostly occur in June, the youngsters typically remain hidden within cover during at least their first week of life, whilst small groups of does and youngsters tend to be seen out in the open from around mid-July. In fact, for much of the year, does and youngsters less than twelve

This fallow buck boasts a fine set of antlers

months old, sometimes in company with yearling bucks, gather in herds separate from those of the older bucks, which in their own groups, wander widely, sometimes straying beyond the boundary of the Crown Lands – only for the annual rut do mature bucks and does regularly associate together.

More...
Fallow deer – preparation for the rut: page 160
The fallow deer rut – all's fair in love and war: page 175
Bolderwood Deer Sanctuary: page 179
Fallow deer – after the rut, bucks feed well and rest: page 200

Great grey shrike – a scarce visitor to Britain

Autumnal and winter visitors from Scandinavia and elsewhere, great grey shrikes prey on small birds and, when available, small mammals, lizards, beetles and other large insects. Large winter territories tend to be occupied from year-to-year, although numbers present in the New Forest can usually be counted in single figures. All normally leave by late March to begin journeys back to the breeding grounds.

Strikingly pale birds with splashes of dark plumage, of similar size to blackbirds and with hook-tipped bills, great grey shrikes are most often first noticed from a distance, perched high in a tree or bush, or on a telephone wire, patiently waiting for the opportunity to take a Dartford warbler or some other unfortunate creature. But roles are sometimes partly reversed by noisily mobbing small birds that vigorously give chase whenever a shrike flies between vantage points.

Yet despite considerable skill as hunters, great grey shrikes, particularly in poor weather, sometimes have difficulty finding food, so in times of surplus, victims' tiny corpses are impaled on gorse or bramble thorns, forming a 'shrike's larder' stocked for use another day.

Great grey shrikes hunt from perches

41

Southern wood ants – tiny members of sophisticated insect communities

Particularly when passing through coniferous woodland, look out for often enormous, dome-shaped southern wood ant nest mounds by the track-side, mounds that when active are home to colonies of ants sometimes tens of thousands strong. However, even though the nests can be over 90 centimetres (35 inches) high and of considerably greater diameter, they blend so well with surrounding vegetation that they often go unnoticed.

Usually placed amongst trees but in fairly open positions that offer the prospect of warming sunshine, these bulky, carefully built and maintained structures generate warmth for the occupants in much the same way as occurs within compost heaps. They are constructed of twigs, leaf stalks and countless conifer needles and may show signs of wear, tear and disturbance by predators, such as badgers and green woodpeckers, seeking a tasty insect meal.

A southern wood ant ventures from the nest

Although during the coldest months, southern wood ants hibernate in underground galleries below the nest, they are active for much of the year and can often be seen on the march across woodland paths: dense, dark columns of insects purposefully going about their business. Up to 10-12 millimetres long, these boldly marked red and black creatures can bite, and spray concentrated formic acid at perceived threats to themselves and their nest, so too close an approach is not recommended.

Along the way

Detours from the main route can provide sightings of a great grey shrike or hen harrier around Ashley Bottom, Ashley Hole, Alderhill Bottom, Ditchend Bottom, Stone Quarry Bottom and Black Gutter Bottom; and on either side of Hampton Ridge.

Small groups of hawfinches, rather scarce, secretive, relatively stout finches with huge bills, can sometimes be seen perched in the upper branches of ash trees around the site of Ashley Lodge, a remote dwelling long ago demolished. (Ash trees, rarely abundant in the New Forest, although the place name here suggests a long history of presence, retain for much of the winter pendulous bunches of winged fruits – ash keys – and it is the single seed within each that, no doubt, attracts the hawfinches.)

Ponies are adept at taking gorse shoots

Fallow deer on sunny days often warm themselves on quiet, open, south-facing valley-sides, such as those encountered around Hampton Ridge.

The Route

1. Leave the Godshill Cricket car park in a south-westerly direction, following the line of the approach road from Brook. Pass a low, Forestry Commission vehicle barrier; go along a wide, grassy track running parallel to the rear of the cricket club pavilion; and continue towards a gap in the gorse 100 metres, or so, ahead.

    Pass a path on the right immediately the gorse is reached, and go straight ahead along a wide, grassy corridor running through the gorse. Ignore paths to left and right; pass beside on the right a moderately steep-sided valley with beyond, distant houses at Godshill; and continue along the now gravel/grassy track as it bears round to the right.

Pass a very minor path on the left; immediately after, go left at a quite substantial junction; and straightaway turn right along a quite narrow, grassy path leading through the gorse. Ignore minor paths on the right and follow the main path as it bears left around the rim of a hill – to the right, the land eventually drops rapidly away towards Forest Brook Farm where a dark coloured, part-corrugated iron barn is prominently visible.

Continue to ignore minor paths on the right until the gorse on the right finishes. Then immediately turn right, downhill, at a crossroad of tracks and head towards a valley-bottom property amongst trees to the left of the farm.

2. Follow the downhill track as it bears left, and turn left alongside a narrow corridor of woodland separating the open Forest from privately owned meadows beyond. Eventually continue round to the right, still alongside the meadows, and cross the Ditchend Brook.

3. Fifteen metres, or so, after crossing the brook, as the main track continues to bear right, beside the meadow edge; go half-left, uphill, along a quite narrow heathland path.

Almost immediately go straight ahead at a relatively indistinct crossroads. After a further 25 metres, or so, turn half-right, still uphill,

Rabbits – timid year-round residents

Rabbits are widely distributed on drier areas of heathland and grassland, and in some woods. Burrows are often excavated on hill-sides and on raised ground provided by Bronze Age barrows. Woodland rabbits, perhaps unwisely, often occupy the same tunnel systems as potentially predatory foxes and badgers, but which of the three are the original owners is usually uncertain. Although rarely occurring in large numbers, and relatively infrequently noticed, rabbits are present throughout the year. (Hares are not often encountered on the Crown Lands, and when present, have often strayed from nearby farmland.)

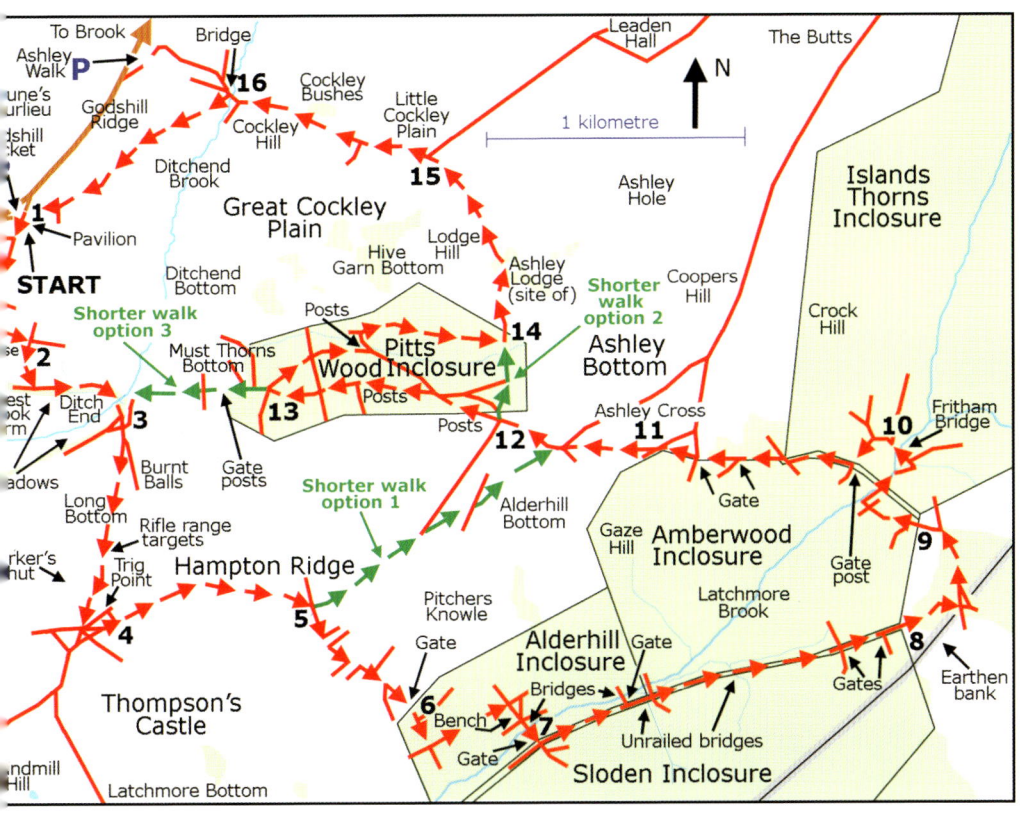

along a relatively inconspicuous though quite wide track leading towards the right-hand end of a promontory some distance ahead – a little to the right of the promontory can be seen a short, conspicuous length of gravel track used later in this section of the walk.

Eventually pass beside the promontory and notice, adjacent to the path, what appears to be a low, concrete shelter.

This is actually part of a target structure originally associated with a volunteers' rifle range established here in 1894. The marker's hut is on the opposite hill-side.

Continue along the now quite wide, gravel track as it goes further uphill, first to the right, and then a little to the left. Reach a six-way junction of tracks close to the hill-top, take the second track on the left and after a short distance, turn left along the gravelled, Hampton Ridge cycle track.

4. Follow the cycle track for around 850 metres to reach a crossroad of tracks – located just after the main track bears slightly right and as it then goes sharply left, up a gentle incline.

5. **To take the first of the shorter walk options** (and miss out the eastern section of the walk), continue along the cycle track and rejoin the main route close to the start of Section 12.

   Otherwise, turn right at the crossroads; continue along a quite wide, initially slightly uphill, gravel track; and follow it as it bears a little left before going down a quite steep hill and on towards the conifers of Alderhill Inclosure.

6. Enter the inclosure through a gate, almost immediately pass a woodland ride on the left, and turn left at the next crossroads to continue along a wide, undulating gravel track.

   Turn right at the next crossroads and go downhill for a short distance along a quite wide, dirt track, past a bench commemorating the life of Eric

Mid-winter birdsong – a precursor to spring

Dunnock

Robins are often paired by early January when their plaintiff autumnal and early winter warble gives way to the stronger tones of spring. Blackbirds from late in the month, at dawn and dusk, quietly, without fuss, announce with song the season's gradually lengthening days. Song thrushes, also primarily at dawn and dusk, call out their repeated series of clear, loud notes; whilst mistle thrushes, the stormcocks of old, exuberantly give voice from the highest, most wind-swept branches. Wrens and dunnocks also sing, and great spotted woodpeckers occasionally drum, but chaffinches usually wait until early February before adding their rollicking notes to the increasing chorus of sound.

Ashby, MBE, a celebrated local wildlife enthusiast and pioneering wildlife cameraman. Immediately after the bench, continue straight ahead, first at a junction with a gravel track, and then at an almost adjacent minor crossroads.

Cross the Latchmore Brook – here flanked on both sides by a nice array of mature oaks and other broadleaves – at a railed bridge.

7. Shortly after crossing the brook, go through a gate and turn left along a wide, grassy driftway between Alderhill and Sloden Inclosures.

After around 400 metres, pass a gate on the left with beyond, a narrow, railed bridge over the brook. Continue straight ahead, after a short distance cross an insubstantial un-railed bridge over a narrow drainage channel, and then immediately pass a gate on the right giving access to Sloden Inclosure and a track on the left leading into here unfenced Amberwood Inclosure.

After a further 300 metres, cross another un-railed bridge over a drainage channel; eventually go straight ahead at a crossroads, with here a gate on the right; continue up a gentle incline and when almost at the edge of the woodland, pass another gate on the right.

8. Leave the inclosure woodland and continue straight ahead through a landscape dominated by ancient, well-spaced, once coppiced/ pollarded hollies. Climb a gentle incline and as the track eventually bears slightly left and then right, go more steeply uphill before

Hazel catkins – early colour in woods and hedgerows

From late January, long before the leaves appear, male hazel catkins – dangling, golden-yellow lambs' tails – conspicuously decorate the trees, incomparably brightening woodlands and waysides. Close examination will also reveal the presence of inconspicuous female catkins: small, swollen bud-like structures with protruding crimson stigmas that superficially look like tufts of short tendrils. Pollination is wind-assisted, a particularly effective strategy when foliage is not present to obstruct the tiny airborne pollen grains.

47

turning left at a crossroad of gravel tracks located just before the hill-top is reached.

A quite pronounced earthen bank is intersected at this crossroads. Running on a north-east/south-west alignment, it is visible from near Whiteshoot Bottom to Watergreen Bottom (both a little outside the area shown on the walk route map) and is associated with a 13th century grant of land to the Lord of Ringwood Manor.

After a short distance, pass a wide, gravel track going back uphill to the right; continue downhill as the track eventually bears left; and when around 100 metres from woodland up ahead, go left at a fairly indistinct 'Y' junction. Almost immediately emerge from the hollies and cross a short stretch of open heathland.

9. Enter Amberwood Inclosure through a gap in the wood-bank – the inclosure here remains unfenced, so there is no gate.

   Almost immediately ignore a grassy track half-to-the-left; continue straight ahead, downhill; go over a crossroad of tracks; and turn right, alongside the Latchmore Brook, keeping the brook immediately on your left-hand side.

   Eventually go right again, alongside a narrow side-stream; almost immediately turn left along a track that crosses the side-stream; and immediately go left again, along a gravelled cycle track that crosses the Latchmore Brook at Fritham Bridge.

Woodlarks – heathland songsters
Cryptically plumaged, primarily heathland relatives of the more familiar skylark, woodlarks are infrequently noticed in early winter, but from late January can be heard singing their beautiful, sustained 'sweet but melancholy' song from the ground, a perch or during slow, circling display flights. Numbers in the New Forest have substantially increased since 1986, when the first national survey of these birds was undertaken, although the skylark population during the same period seems to have declined.

10. Follow the cycle track as it bears left, passes through a gap in the woodbank separating Islands Thorns Inclosure from Amberwood Inclosure – marked on the left by a single gate post – and then goes right.

    Eventually leave behind Islands Thorns Inclosure and continue uphill with Amberwood Inclosure to the left; and to the right, a more open landscape of bracken, hollies, occasional oaks and Scots pines.

    Ignore a gated gravel track on the left, continue up a quite steep hill and after a short distance, close to the top of the hill, go straight ahead along the cycle track where to the left is a gate and path leading into the wood, and to the right, a gravel track.

11. Follow the cycle track out onto the heath, eventually go straight ahead at a junction of minor tracks and immediately after, leave the cycle track by taking the right fork at a 'Y' junction to continue along an equally wide, gravel track.

    Ignore minor paths on the right as the main gravel track bears left, downhill; and after a short distance, reach a crossroads.

12. **To take the second of the shorter walk options** (and avoid much of Pitts Wood Inclosure), turn right here, go downhill through the edge of the wood, cross a narrow stream and rejoin the main route at the start of Section 14. (An iron plate outlining the history of inclosure can be found on the right as the wood is entered.)

    Otherwise, go straight ahead at the crossroads, still downhill; and enter the inclosure between two solid old gate posts. Pass a minor path on the right, immediately take the left fork at a prominent 'Y' junction and eventually pass three more, substantial posts. Ignore a path going left, uphill, immediately after the posts; continue downhill and then go straight ahead at a crossroads.

13. **To take the third of the shorter walk options** (and miss out much of the northern part of the route), go straight ahead at the next crossroads – just before the edge of the wood is reached – and continue downhill. After a short distance, pass on the right, a quite wide track; and after a further short distance downhill, go between two substantial old gate posts to leave the woodland behind – on the left here is another old, iron inclosure plate. Immediately go straight ahead at a crossroads of wide grassy tracks, and continue straight ahead across the heath to reach Ditch End at the stream crossing encountered at the end of Section 2. From there, retrace the outward route back to the car park.

49

Otherwise, turn sharply right at the next crossroads; go straight ahead at another crossroads; and shortly after, pass through a grove of aged hollies and oaks, some of which almost certainly pre-date the original inclosure planting.

Pass between two more, sturdy gate posts; immediately after, turn left at a 'T' junction; continue downhill along a gravel track and cross a usually narrow stream. Ignore a minor track on the left and follow the gravel track as it bears round to the right.

14. Turn left at the next 'T' junction of gravel tracks and continue uphill, eventually past the site of Ashley Lodge with its associated, somewhat uncharacteristic for the New Forest, tangles of hawthorns, blackthorns and brambles that provide a strong, wildlife-rich under-storey.

    Keep to the main gravel track, pass through an area of hollies and reach open ground close to the top of the hill.

15. At Little Cockley Plain, pass another gravel track on the right leading to Leaden Hall; continue straight ahead, eventually quite steeply downhill; and pass on the right, Cockley Bushes, a stand of hollies and other trees much favoured by winter thrushes as a source of berries and shelter.

16. Cross in the valley-bottom, a bridge over the Ditchend Brook and immediately turn half-left, off the gravel track, to follow a quite pronounced track running diagonally uphill over the heath – after crossing the bridge, ignore another track running uphill, quite close to the main gravel track.

    When almost at the top of the hill, ignore a minor track on the left and continue along the more substantial main track as it bears round to the right.

    After a short distance, emerge from the gorse and immediately ahead are the cricket pitch, pavilion and car park.

For the adventurous, for those with a good sense of direction, strong map reading skills and access to an Ordnance Survey map!
Create your own walk by combining parts of this route with elements of your choice from the following selection of connecting or conveniently located nearby routes.

From *New Forest Walks – a seasonal wildlife guide:*
    August   Ashley Walk, Millersford Bottom

From *New Forest Walks – a time traveller's guide:*
  Walk 1   Ashley Walk Bombing Range
  Walk 2   Bramshaw Telegraph
  Walk 16   Frogham: Latchmore Brook

Start	Busketts Lawn, Forestry Commission car park, near Bartley Cricket Club's pitch, 4 kilometres (2½ miles) south-east of Cadnam. Ordnance Survey map reference SU311112
Distance	8.5 kilometres (5¼ miles) Shorter walk option: Reduce the distance by 2.25 km (1½ miles)
Time to allow	2 - 5¼ hours
Refreshments	The New Forest and the Happy Cheese, Ashurst; the Gamekeeper, Woodlands Road, Ashurst; and the Haywain, Bartley are all relatively nearby
Route	Mainly along readily visible tracks, but in a small number of places – for example, during Sections 8 and 9 – a little 'off the beaten track'
Terrain	Mainly on level ground, although a small number of relatively gentle gradients are present. Be aware, too, that, particularly in winter and after heavy rain, crossing Bartley Water at a ford encountered during Section 17 can be very awkward without Wellington or other tall, waterproof boots
Rating	2 – moderate walking
Buggies	Not suitable
Railway station	Ashurst (New Forest), 3 kilometres (1¾ miles)
Bus service	Bluestar
New Forest Tour Bus	Yes
Alternative start	Brockishill Green, Forestry Commission car park at Ordnance Survey map reference SU299117
'Camping in the Forest' Caravan and Campsite	Ashurst, 3.25 kilometres (2 miles)

FEBRUARY
Welcome signs of spring
(Bartley Cricket, Busketts Lawn, Furzy Lawn Inclosure, Fox Hill, Rushpole Wood and Busketts Lawn Inclosure)

Early morning mist in winter woodland

The walk
Starting at Busketts Lawn car park, adjacent to the ground of Bartley Cricket Club, this 8.5 kilometre (5¼ mile) primarily woodland walk also passes over heathlands and grasslands, and for some of the way follows beside Bartley Water, an attractive, meandering New Forest stream. A short-cut is available to reduce the distance by 2.25 kilometres (1½ miles).

Featured wildlife

Grey squirrels – spring is in the air

In years when acorns and beech-mast are abundantly produced, grey squirrels will feed well throughout the autumn and much of winter, then if in the rudest of health may mate in February. Yet if food has been hard to find and the animals are in less than optimum condition, mating will be delayed until spring or summer. But whatever the timing, the squirrels' amorous, energetic, treetop pursuits invariably attract attention as, temporarily oblivious to watchers below, these much maligned creatures provide impressive demonstrations of speed, balance and agility, illustrating absolute command of their precarious environment high above the ground.

Up to 60 centimetres (24") across and constructed of twigs wedged into a convenient tree fork, litter dreys are lined with moss, grass, leaves and whatever other soft, comfortable material is available. Alternatively, dens made in tree cavities may be used, much to the chagrin of jackdaws, stock doves and other birds displaced from nest or roost sites.

A grey squirrel surveys its domain

The squirrels' gestation period is around forty-five days so, depending upon the time of mating, births can occur in spring, summer or early autumn. Litter size averages three, but can be anything from one to eight. Youngsters – known as kittens – are weaned at ten weeks and become independent at ten to sixteen weeks.

More...
Grey squirrels – widespread New Forest colonists: page 194

Holly – an attractive, prickly evergreen

Although widespread as an under-storey tree of modest stature and a regular heathland inhabitant, the humble holly often goes unnoticed. Yet the presence of aged multiple stems eloquently betrays past coppicing and pollarding, both formerly widespread practices that encouraged the growth of fresh new shoots, for many of the older hollies have been extensively used by past generations of Forest people: charcoal burners and others harvested the timber to fuel their fires, bird-catchers produced birdlime from the bark, and keepers and commoners valued the leaves and twigs for use as browse for deer and ponies.

Indeed, holly continues to be cut in winter for the benefit of deer and stock, whilst pollarding has been resumed so as to rejuvenate the trees, prompt the appearance of new shoots and allow more light to reach the woodland floor which, in turn, encourages the growth of wild flowers and other vegetation.

Holly berries have usually all been taken by February

But not only do the leaves – especially prickly low in the tree, less so higher up – and twigs help satisfy the appetites of deer and ponies, for, particularly in late winter when other foodstuffs are in short supply, the bark is also often nibbled, producing characteristic tell-tale scars at heights that reflect the animals' reach.

Both male and female hollies in late spring and early summer bear clusters of tiny white flowers that when dislodged from the trees, may envelope the ground, snow-like, beneath the boughs. Only females, though, bear the berries that glow red in late autumn and early winter, and are so enjoyed by blackbirds, fieldfares, mistle thrushes, redwings and song thrushes. However, in common with oak, beech and many other trees, the size of the berry crop is extremely variable and is rarely strong in consecutive years.

Ivy – woodman spare that axe

An evergreen, woody climber capable of prolific growth, ivy provides valuable services to a wide range of wildlife. To the delight of many insects, its greenish-yellow flowers are at their best in autumn, when few other sources of pollen and nectar are available; whilst in February and March, clusters of black fruits offer late winter foodstuffs appreciated by birds and small mammals. Birds also nest amongst the stems, holly blue butterfly caterpillars in spring and summer gorge on the leaves and the dense foliage provides year-round shelter for a wide variety of species.

But by competing for nutrients and increasing the likelihood of wind-throw damage, ivy was believed to cause undue harm to the trees on which it grew, and so its stems were ruthlessly cut; but now, in these somewhat more enlightened days, the plant is often left to flourish.

Kingfishers – colourful birds of streams and rivers

Kingfishers are present on many New Forest streams, particularly from late February until the end of summer, but frequently go unnoticed – their small size and relatively secretive habits help the birds to avoid detection, although shrill, piping calls offer clues to their whereabouts.

Flight, on rapidly whirring wings, is usually low over the water, but kingfishers sometimes avoid stream meanders by flying through adjacent trees, whilst excursions are occasionally enjoyed above the treetops or over open land. Steep-banked woodland streams are favoured as breeding sites, but these birds may be found alongside virtually any stretch of water, especially early in their period of residence as newcomers travel in from coastal and other wintering grounds, and again later as youngsters of the year and adults move away from the Crown Lands.

After the eggs have hatched, occupied nest sites are sometimes

Ivy berries provide late winter feasts for blackbirds and other creatures

A kingfisher waits patiently beside a stream as it looks out for fish in the water below

betrayed by the unedifying spectacle of droppings oozing from tunnel entrances, droppings that subsequently may encourage the growth of vegetation in the enriched sand or soil. Close approach by walkers and others at this time is clearly not recommended. Indeed, any steep, vertical, streamside bank is best kept at a respectful distance in spring and early summer lest a pair of breeding kingfishers be disturbed.

Lichens – fascinating, often overlooked members of a miniature world

Lichens, unlikely associations of fungi and algae, are prominently visible in the bare February woods. Almost 300 species have been recorded growing on bark or exposed wood in the New Forest, including some that are extremely scarce elsewhere. Many have poor powers of dispersal and are consequently considered to be indicators of ancient woodland where tree cover has been continuously present for a considerable period of time. And somewhat reassuringly, many are also indicators of the presence of clean, unpolluted air.

Busketts Wood, close to the start of this walk, is particularly lichen-rich – 174 species per square kilometre were discovered there during surveys undertaken up to 1983. Lichens are also impressively present along the route through Rushpole Wood, whilst heathland, such as that around Fair Cross, is host to a further range of specimens that often grow in bare patches of ground amongst the taller vegetation.

Conspicuous examples amongst many much less so include members of the *Cladonia* genus – branch or twig-like lichens such as *Cladonia diversa*, a stalked species with red fruiting body, found on rotted trees and in acid soils. Look out, too, for the similarly stalked but greenish-grey *Cladonia chlorophaea*, a lichen that grows in soil, on rotting stumps and on mossy banks; and for *Cladonia portentosa*, a heathland inhabitant found amongst the

Lichens festoon a fallen beech tree

heathers. Beard lichens of the *Usnea* – rope-like – genus, such as the grey-green *Usnea ceratina*, are also relatively distinctive and widespread.

Roe deer – timid creatures of wood and heath

The smallest of the commonly seen New Forest deer – height at the shoulder is up to 73 centimetres (29 inches) – roe deer numbers are currently maintained at a minimum level of around 400 animals. Although a native of the British Isles, roe deer had probably been hunted to extinction in the New Forest by the end of the 16th century and did not re-colonise until the second half of the 19th century.

Despite their widespread local distribution, many roe deer tend to gravitate towards Crown Land boundaries where better feeding is available in nearby meadows, privately owned woodlands and residential gardens. They are most

A roe buck, its antlers in velvet, crosses a Forest track

often seen singly or in small groups of two or three – bucks and does, for example, are most often seen together during the build up to and duration of the annual rut, whilst does and their most recently born youngster frequently remain together until close to the time of the next year's birth.

Summer coat is bright reddish-brown; winter coat is dark greyish-brown above and paler below. Spots are absent at all times, apart from on newborn roe deer. Some animals as early as February show traces of reddish-brown around the back of the neck as the moult into summer coat begins, but this is not usually fully underway until April. Moult into winter coat starts in the second half of September or in October. Their black nose, bordered by small pale patches and pale chin, is distinctive at all times, and so is their pale, powder-puff rump patch.

Antlers are quite short, round in section, fairly straight and usually have three points on each. Old headgear is cast between mid-November and mid-December, re-growth quickly starts and the new antlers remain 'in velvet' until April – removal of the velvet is often assisted by vigorous rubbing against young trees and branches, which can cause considerable forestry damage.

More...
Roe deer – summer rut and spring births: page 128

Snipe – inhabitants of swamps and other wet ground

Secretive birds of the New Forest's bogs and mires, snipe thrive in damp conditions where their lengthy bills enable invertebrate foodstuffs to be plucked from deep in the mud. Present in varying numbers all-the-year-round, they are blessed with cryptically patterned plumage – an attractive mosaic of streaky browns combined with darker and lighter shades – that helps them avoid detection when on the ground. In fact, for much of the year, other than when disturbance prompts harsh alarm calls and rapid, zigzag escape flight, they are often overlooked altogether.

In spring, however, attention is

A snipe stretches its wings

attracted by bouts of breeding season calling – quite loud, repeated *chipper-chipper-chipper* notes – from within dense cover, and primarily early morning and evening displays during which one or more males chase about the sky, often almost lost to sight high above, intermittently plunging earthwards with tail fanned and the outermost feather on each side held out to produce a distinctive humming sound known as drumming or bleating.

Snipe from early April typically lay four eggs in a well-concealed nest on the ground and in common with other ground nesters, such as curlews, lapwings and redshanks, are particularly susceptible to disturbance by people and dogs that encroach too closely on their space.

More...
Breeding waders – snipe, curlews, lapwings and redshanks: page 94

Along the way

An attractive, typically narrow New Forest stream, Bartley Water is first encountered during Section 8 of this walk and again during Sections 15, 16 and 17. Mallards and mandarin ducks are likely to be present all-the-year-round and so, too, are grey herons, large, ungainly birds that feed in shallow water, tucked away out of sight alongside quiet meanders and in other places

Bartley Water in early spring

little visited by people. Grey wagtails and kingfishers also live beside the stream, but are most likely to be seen in spring and early summer, whilst little egrets, diminutive relatives of the grey herons and relatively recent UK colonists, are primarily present in autumn and winter.

From late May until late August, above the water's surface, metallic blue-green male beautiful demoiselle damselflies engage in extravagant dancing display flights intended to impress nearby, primarily golden brown females, whilst a range of other dragonflies and damselflies might also be seen.

The Route

1. Leave the Busketts Lawn car park at its south-western corner – the corner closest to Bartley Cricket Club's pitch. Pass a low, Forestry Commission vehicle barrier and continue along a gravel track leading in the direction of the pitch.

    After a short distance, leave the gravel track and continue between the cricket pitch and adjacent pavilion.

2. When almost at the far end of the cricket pitch fence, turn left and walk along the left-hand edge of a quite wide, grassland corridor from which trees have relatively recently been removed.

    After around 350 metres, reach a broader, more extensive area of grassland running at around 90 degrees to the original grassland corridor; and go to the right, across the open neck of that original corridor. (On the left just before this turn, four railway sleepers (or similar planks of wood) form a makeshift bridge over an often wet patch of ground.)

3. Skirt a narrow area of woodland on the right and almost

Unseasonal butterfly appearances
Brimstones, commas, peacocks and small tortoiseshell butterflies regularly over-winter as adults, hidden away in garden sheds, hollow trees and amongst clumps of ivy; and sometimes take to the wing on bright February days, adding welcome splashes of colour to the countryside.

Brimstone

61

immediately follow its edge round to the right, alongside another grassland corridor.

4. Ignore a track on the right leading into the wood, and instead, enter the woodland using another track, at the far left-hand corner of the grassland corridor.
    Continue up a gentle gradient; when almost at the top, go straight ahead at a crossroad of tracks; pass a hill-top track on the right; go downhill for a relatively short distance; emerge into a small clearing and go through a gate on the far side of this clearing to enter the primarily coniferous woodland of Brockishill Inclosure.

5. Go straight ahead along a quite wide, grassy woodland ride; pass a similar ride on the right, and minor paths to left and right – those on the left lead to nearby Dogben Gutter, a narrow stream that eventually joins Bartley Water.

6. Turn left at the next crossroads to follow along a gravelled cycle track. Shortly after, pass into Furzy Lawn Inclosure – there is no gate or fence here. Continue along the cycle track and go straight ahead at the next crossroads.

7. After a relatively short distance, turn left at a further crossroads (where the gravel cycle track is intersected by a grassy ride). Soon after a right-hand bend, take the right-hand fork at a grassy 'Y' junction and go through a gate into an area of ancient, unenclosed woodland.

8. Go straight ahead, downhill, and after a short distance use a railed bridge to cross Bartley Water.
    Caution – take particular care during the remainder of this section as a number of the paths are relatively indistinct.

    Follow the path as it zigzags uphill. When around 40 metres from the bridge, pass a narrow, rather inconspicuous path on the left and continue uphill along the wider path.
    After another 40 metres, as the ground levels out, pass another inconspicuous path on the left. After another 20 metres, go right at an indistinct 'T' junction. (The open ground of Gutter Heath is visible through the trees to the left here.)
    Follow this path as it soon bears a little to the left. When around 70 metres from the previous 'T' junction, emerge from the woodland onto the

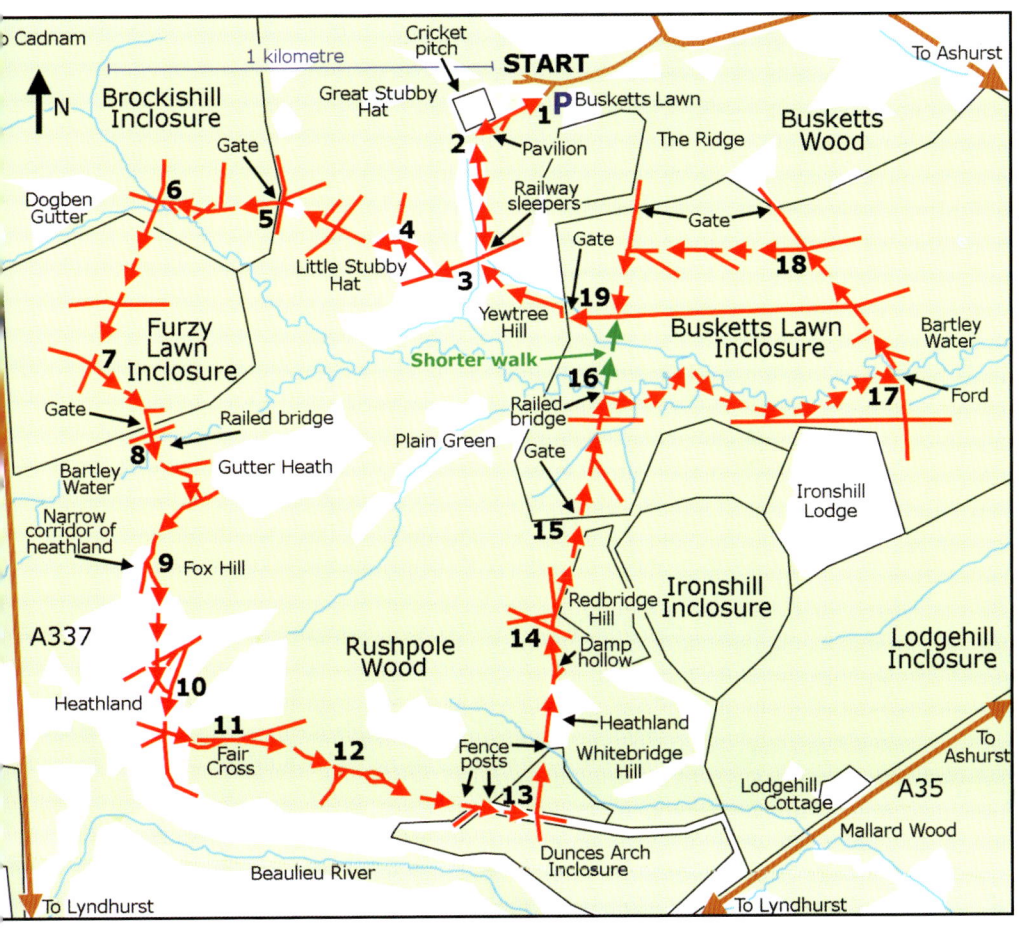

edge of a narrow corridor of heathland adjacent to the wooded rising ground of Fox Hill, which is on the left.

9. Follow a narrow, woodland edge path – initially running half-to-the-right for a short distance – around the middle contours of the hill, avoiding here another path further to the right, running around the base of the hill.

After around 250 metres, follow the path as it leaves the wooded hillside and strikes off straight ahead across the heath. Ignore minor tracks to left and right, and when 150 metres, or so, beyond the hill-side, and almost adjacent to a line of trees on the left, go left for a very short distance through the heathland vegetation.

10. Almost immediately turn right along a wider, more pronounced track, and after a relatively short distance, at a 5-way junction of tracks, turn left along a quite narrow, grassy path leading uphill into Rushpole Wood.

Almost immediately ignore a minor path on the right; continue along the main path as it bends slightly left through the wood, quite close to heathland edge on the left; and after a further relatively short distance, ignore another minor path on the right.

11. Go straight ahead, almost immediately pass a left-hand fork and carry on, following the undulating path's numerous meanders and ignoring along the way, minor paths to left and right.

Gorse – colourfully widespread furze

At their dazzling best in late winter and early spring, gorse blooms at this time bathe the heaths in a cloak of golden yellow, although the old say 'kissing is out of fashion when gorse is not in bloom' indicates that these sometimes quite large, sturdy bushes bear at least some flowers throughout the year.

Mature, often almost impenetrable, gorse brakes are much loved by commoners' ponies for the shelter they provide and also as an oft taken source of unlikely yet appetising food. Linnets nest amongst the branches, stonechats use the outer reaches as perches from which to admonish passers-by and Dartford warblers habitually hunt within the tangles for spiders and insect foodstuffs.

Local people, too, appreciated gorse for use as fodder for the stock – it was often cut and crushed to improve palatability – whilst faggots were used as fuel for the fire.

Gorse blooms are at their best in late winter and early spring

More....
Dwarf gorse – heathland ankle nipper: page 154

12. Eventually pass a narrow track on the right and after a short distance, follow one of the two paths that run either side of an obstructing clump of holly, birch and other trees.

    Continue straight ahead, eventually downhill; pass a quite large fence post on the left; a wide, grassy ride on the right; and another fence post on the left. Again go straight ahead, leaving behind the broadleaved trees of Rushpole Wood to walk along another quite wide, grassy ride bordered by birch trees and conifers.

13. After a short distance, when almost at the top of a gentle incline, turn sharp left at a crossroads and go downhill through the wood, eventually past a number of old gate/fence posts and out into an area of heathland onto which bracken, beech trees, oaks and hollies are encroaching.

    Continue uphill and when across the brow of the hill, plunge back into Rushpole Wood. Immediately ignore an indistinct fork to the right and continue straight ahead up a hill of modest gradient – it is often necessary here to skirt a small, damp hollow.

14. When almost at the top of the hill, go straight ahead at a crossroads; almost immediately ignore a very indistinct track on the left (opposite a Rising Main marker); again almost immediately, take a relatively indistinct right-hand fork at a 'Y' junction and continue downhill.

Goshawks – time for sudden, often unexpected appearances

Considering their relatively large size, similar to that of buzzards, goshawks in woodland frequently evade observation by sitting still, high in the canopy or else when on the wing, by silent, direct flight through the trees. But primarily in late winter and early spring, members of the modest sized but increasingly successful New Forest population for a short time become conspicuous, for that is when they engage in spectacular aerial displays. Look out, for example, for 'sky-dancing', a series of manoeuvres performed mainly by females during which the birds soar upwards in tight spirals before plunging headlong into the breeding woods.

Goshawks are ferocious predators: wood pigeons are favourite prey items, whilst grey squirrels will also be taken

65

15. After around 250 metres, go through a gate into Busketts Lawn Inclosure; join a gravel cycle track at an inverted 'Y' junction and go straight ahead.
    Go straight ahead again, at a crossroads where the cycle track intersects a wide, grassy ride, and reach Bartley Water at a railed bridge.

16. **To take the shorter walk route**, which misses out much of Busketts Lawn Inclosure and a Bartley Water ford (which is often difficult to cross without Wellington or other tall, waterproof boots), continue straight ahead here; cross the bridge; and at the next crossroads, where the cycle track turns sharply right, go left and follow the walk directions from Section 19.

    Otherwise, to enjoy a river-side walk below majestic Douglas firs, turn right immediately before the bridge and follow a narrow path beside the stream. After a very short distance, pass to the left of a wide, overgrown, largely disconnected meander cut off from the main flow and continue to follow the stream's every twist and turn, leaving the waterside only as the path's course dictates.

Tawny owls – day-time roost sites betrayed
Numerous and widespread woodland dwellers, tawny owls, as if quietly murmuring in their sleep, occasionally hoot during the day. Presence, though, is more likely to be revealed by angry throngs of mobbing smaller birds – jays, blackbirds, chaffinches and others – that have found an owl's hideout deep in a holly, amongst ivy or on a branch close to the trunk of a tall tree; and whilst always maintaining a safe distance between themselves and their target, are anxious to draw attention to the predator and force flight.

Eventually pass beside on the right, a grassy woodland ride with beyond, the grounds of Ironshill Lodge.

17. Reach a gravel track close to the Bartley Water ford, turn left and cross the stream at the ford.
    After a short distance, pass a woodland path on the right, and shortly after that, another on the left; and then go straight ahead where the walk route crosses a gravel cycle track and continues as a wide, grassy ride.

18. Turn left at the next crossroads and continue down another grassy ride. Pass two quite wide, yet relatively indistinct grassy rides on the left, and immediately after the second, turn left at a 'T' junction along yet another grassy ride.
    Turn right at the next crossroads – the gravel cycle track used during Section 15 forms two of the branches here.

19. Continue along this wide, grassy ride; leave the inclosure through a gate and immediately turn right, alongside the fence-line. Then almost immediately go left, parallel to a narrow, sometimes dry, drainage channel bordering an area of grassland interspersed with bracken and mature trees.
    Follow this drainage channel as, after around 200 metres, it bears to the right and meets the first grassland corridor used on the outward route – close to the railway sleepers referred to earlier.
    Retrace your steps back to the cricket pitch and car park.

For the adventurous, for those with a good sense of direction, strong map reading skills and access to an Ordnance Survey map!
Create your own walk by combining parts of this route with elements of your choice from the following selection of connecting or conveniently located nearby routes.

From *New Forest Walks – a time traveller's guide:*
   Walk 5   Lyndhurst Golf Course and the old Race Ground
   Walk 6   Ashurst: Busketts Inclosure

Start	Tilery Road, Forestry Commission car park, 1.5 kilometres (1 mile) north-east of Brockenhurst village centre, off the B3055 Brockenhurst to Beaulieu road – Ordnance Survey map reference SU308033
Distance	7.5 kilometres (4½ miles) Shorter walk option: reduce the distance by 3 km (1¾ miles) Walk extension: a detour into Frame Heath Inclosure provides the potential to extend the route
Time to allow	1¾ - 4½ hours
Refreshments	The Snakecatcher, the Rose and Crown and the Foresters Arms are all in Brockenhurst village
Route	Along gravel tracks
Terrain	Mainly on level ground
Rating	2 – moderate walking
Buggies	Suitable for sturdy buggies
Railway station	Brockenhurst, 1.75 kilometres (1 mile)
Bus service	Bluestar and National Express
New Forest Tour Bus	Yes
Alternative starts	1) Standing Hat, Forestry Commission car park at Ordnance Survey map reference SU314036 2) Balmer Lawn, Forestry Commission car park at Ordnance Survey map reference SU303031
'Camping in the Forest' Caravan and Campsites	1) Hollands Wood, 0.5 kilometres (⅓ mile) 2) Aldridge Hill, 4 kilometres (2½ miles)

MARCH

In like a lion, out like a lamb?

(Brockenhurst: Balmer Lawn; Standing Hat; Pignalhill, New Copse, Frame Heath, Perrywood Haseley, Parkhill and Pignal Inclosures)

Balmer Lawn

The walk

A walk in the woods near Brockenhurst is always an attractive proposition. Starting at the Tilery Road car park, adjacent to Balmer Lawn, this 7.5 kilometre (4½ mile) route is mostly on level ground. A short-cut is available to reduce the distance by 3 kilometres (1¾ miles), whilst a detour into Frame Heath Inclosure, where there is always a chance of seeing sika deer, offers potential for extension.

Featured wildlife

Goldfinches – splendidly colourful and surprisingly confiding

The goldfinch is surely one of our most attractive birds. Certainly the Victorians thought so, for they caught large numbers for display in cages – in 1860 alone, 132,000 are said to have been caught on the South Downs near Worthing. In fact, the significant depletion of wild populations was one of the early concerns of the fledgling RSPB following its formation in 1889. Live trapping is now outlawed in the UK, but such is the popularity of these birds that illegal capture remains a problem.

Goldfinches display clearly visible patches of red about the face

Goldfinch! The name is absolutely appropriate, for both males and females display broad, strikingly golden-yellow wing bars that are clearly visible in flight and at rest. But redcap, one of many old, country names, is almost as descriptive, although good views reveal that the birds also have red about the face, rather than just on the cap. The Anglo-Saxons of the 8th century, however, knew the bird as Thisteltuige, or thistle-tweaker, a reference to its insatiable appetite for thistle seeds.

Although widespread in much of the Hampshire countryside, goldfinches are relatively infrequently seen on the Crown Lands of the New Forest and have probably never been common. Yet flocks, delightfully known as charms, are sometimes encountered and the birds' tinkling, musical calls heard, not only in summer and early autumn in places where thistle seeds and other foodstuffs are available, but also in spring as travellers return from winter wanderings that sometimes take them as far south as Spain.

Honeysuckle – the promise of extravagant blooms to come

One of the first plants to produce new leaves in the early year, honeysuckle is blessed with exotic, creamy-yellow blooms, often tinged with lilac, red or orange, that can be seen from mid-May through to October. Their rich, sweet scent is particularly noticeable at dusk and through the night when a range of moths are attracted to feed on the nectar, and whilst so doing, pollinate the flowers. Subsequently produced bright red berries, whilst of handsome appearance, are poisonous to humans, but are readily taken by bank voles, blackbirds, blackcaps and a host of other creatures.

Honeysuckle blooms are particularly attractive to night-flying insects

A robust climber ever motivated to seek out the best of the light, honeysuckle clings so tightly to its host that sapling trunks are sometimes squeezed into spiral shape. It grows well in hedgerows, thickets and open woodlands, but is also found along dappled woodland rides and in other relatively shaded places where thin, spindly examples are the sole foodplants of white admiral caterpillars. Indeed, without honeysuckle, this handsome butterfly would be entirely absent from our woods.

Magnificent oaks – so important for wildlife

Trees of history and tradition, long-lived, sacred and venerated; oaks are solid, reliable, durable and drought-resistant. Pedunculate oaks – sometimes referred to as common oaks or English oaks – live side-by-side with closely related but, in the New Forest, less frequently seen sessile oaks. (In contrast to sessile oaks, pedunculate oak leaves have virtually no stalk, whilst their acorns are borne on stalks an inch or two long. There is, however, the potential for confusion caused by the presence of a variety of related species and hybrids.)

Red oaks and Turkey oaks, trees introduced to Britain in the early 18th and second half of the 19th century, respectively, were also found in the New Forest, albeit in relatively small numbers, but many, if not all, have in recent years been destroyed as part of a programme devised to remove specific exotic and pest species. Yet despite the loss of these and countless other oaks cut over the centuries to provide navy ship building timber, many fine, aged specimens continue to withstand winter gales and the depredations of ravenous fungi. The oldest – some probably more than 400 years of age – usually occur in the ancient, unenclosed woodlands, although some 19th century, and earlier, inclosures – such as New Copse and Parkhill Inclosures,

The branches of this ancient pollard oak are mostly dead and decaying, but some continue to bear small numbers of leaves

both passed through during this walk – contain oaks that date from the time of original enclosure.

In spring, conspicuous male catkins – lengthy, hanging stalks bearing clusters of vivid yellow-green flowers – can be seen on the same trees as the much less noticeable females. Pollination is by the wind, and the resultant acorns, sitting comfortably within supporting cups, fall in early autumn. The strength of the crop is, though, extremely variable from year to year, with production stimulating 'feast' or 'famine' conditions for the many creatures that depend upon acorns for autumnal and winter foodstuffs – remarkably, it has been calculated that in times of plenty (reckoned to occur every three or four years), a large, productive tree can yield up to 50,000 acorns.

A staggering variety of wildlife species – birds, mammals, insects and other invertebrates, fungi, flowers, mosses, lichens – live in and upon oak trees. In fact, pedunculate oaks are said to support a greater variety of species than any other type of tree in the British Isles.

More...
The mast crop – a substantial influence on the well-being of woodland wildlife: page 170

Sika deer – mysterious shapes in the woods

Sika deer, singly or in small groups, may be encountered during this March walk in and around New Copse and Frame Heath Inclosures. Originally imported from Asia onto the Beaulieu Estate at the turn of the 19th century, sika deer numbers have in recent years been maintained at a maximum of around one hundred animals whose distribution is largely restricted to the woodlands and adjacent heathlands south of the railway line, to the east and south-east of Brockenhurst. (Revised policy outlined in the South England District Deer Strategy, 2014-2020 proposes a reduction in herd size to no more than seventy animals).

A sika stag, ever alert to potential danger

Although closely related to red deer, sika deer are more similar in size to fallow deer. Summer coat is chestnut-red or fawn, with distinct pale spots; but in thick, dense winter coat, stags can appear almost black and seem to carry a perpetual frown, whilst hinds in winter coat are greyish-brown in colour. Moult into summer coat starts in spring, and into winter coat in autumn. Both sexes can be reasonably readily identified by the presence of a relatively large, white, heart-shaped rump patch with black surrounds and a thin, dark line down the primarily white tail.

Multi-branched antlers on mature stags are round in section and usually carry four points on each. They are typically cast in April or May and re-growth starts almost immediately. Prior to and during the rut, as a means of advertising mating territory, these formidable weapons are used to scour deep, often twin, vertical scars into tree trunks, providing deer enthusiasts with year-round clues to presence.

Sika deer start to rut in mid- to late September in a mating process that can sometimes continue, eventually at low levels of intensity, until December.

Rutting calls made by the stags are particularly distinctive: horrible, shrill shrieks and high-pitched whistles, often repeated three to five times in quick succession. Births, normally of a single calf, commonly take place from mid-May to mid-June, although as the rutting period is somewhat extended, births can occur from early May right the way through to September or October.

Willow trees and bumblebees – a timeless combination

One of the most conspicuous signs of spring, male goat willow – pussy willow – catkins glow golden in the still-weak sun; whilst on separate trees, less flamboyant, silvery females go largely unnoticed. Both sexes, however, are irresistibly attractive to a variety of insects at a time when few other pollen and nectar sources are available. Eventually, though, their work over for another year, the male catkins' splendour fades as the females go on to produce distinctive woolly seeds attached to long, white hairs that assist distribution by the breeze.

A tree bumblebee forages for pollen on a willow catkin

Bumblebees, heads and bodies dusted bright yellow with pollen, are frequently present guests at the catkin feast. Attractive, rotund, often understated creatures, queen bumblebees freshly woken from hibernation are early visitors, but they will largely retire to the nest later in spring or in early summer when newly hatched, smaller female worker bees take on the task of collecting food for the growing colony. Some species are particularly well-equipped with lengthy tongues that give ready access to the catkins' tiny, well-hidden nectaries, although only females have leg adaptions, aptly known as 'pollen baskets' – flat, shiny areas fringed with stiff hairs, located on the outer surface of each hind leg – that enable dustings of pollen moistened with nectar to be stored and flown back to the nest.

Males and new queens hatch later in the year, feed and seek mating opportunities. The males do not collect food for the colony, nor until the following year do the young queens. In fact, it is only these young queens that survive the winter, safely hidden away in the soil where they hibernate alone until spring. After emergence, they will feed on nectar and pollen; provision a new nest – located underground, for example, or in a grass tussock, disused birds' nest or tree hole – also with nectar and pollen; secrete wax from their bodies and start the annual cycle afresh by laying eggs in an untidy cell (made from wax and pollen) before going on to raise a first brood of female workers.

Twelve bumblebee species can be found in the New Forest, although only six are relatively common and widespread: the garden bumblebee, red-tailed bumblebee, white-tailed bumblebee, common carder bumblebee, early bumblebee and the buff-tailed bumblebee.

Along the way

Often dismissed as of little value for wildlife, coniferous woodland, such as that encountered during this walk, really does get a bad press, but there is always some life amongst the dark foliage and beanpole trunks, particularly if a scattering of broadleaved trees are also present.

Coal tits, common crossbills, goldcrests and siskins, for example, may all be present. Firecrests, absolutely notable birds, might also be seen – the first UK breeding season reports for the species came from Bolderwood in the early 1960s – whilst tawny owls and occasional long-eared owls find sheltered daytime sanctuary on branches shrouded in evergreen foliage. Conifers also

This roe buck is reluctant to leave its sanctuary within primarily coniferous woodland

feature prominently in the lives of some of our larger, more spectacular birds, for buzzards, goshawks, hobbies, kestrels, ravens and sparrowhawks all use these trees as nest sites, in some cases to the virtual exclusion of all others.

Deer and foxes, too, appreciate the potential for concealment provided by coniferous woodland and frequently lie-up amongst the trees during the day before emerging at dusk to feed along the rides, in nearby broadleaved woodlands and on the heaths.

The Route

From Brockenhurst village centre and the railway station

Leave the village centre on the main A337, heading north towards Lyndhurst. When 1 kilometre (0.6 miles) beyond the Brockenhurst railway level crossing, go over a river at a road bridge and immediately after, turn right into Balmer Lawn Road, the B3055 signposted to Beaulieu and Roundhill Campsite. (The Balmer Lawn Hotel can be seen to the left, whilst the Balmer Lawn, Forestry Commission car park is beside the river on the right.)

After 400 metres, where the road bends to the right, follow a wide, gravel Forestry Commission track leading straight ahead towards the distant woods. The Tilery Road car park is a short distance ahead on the left.

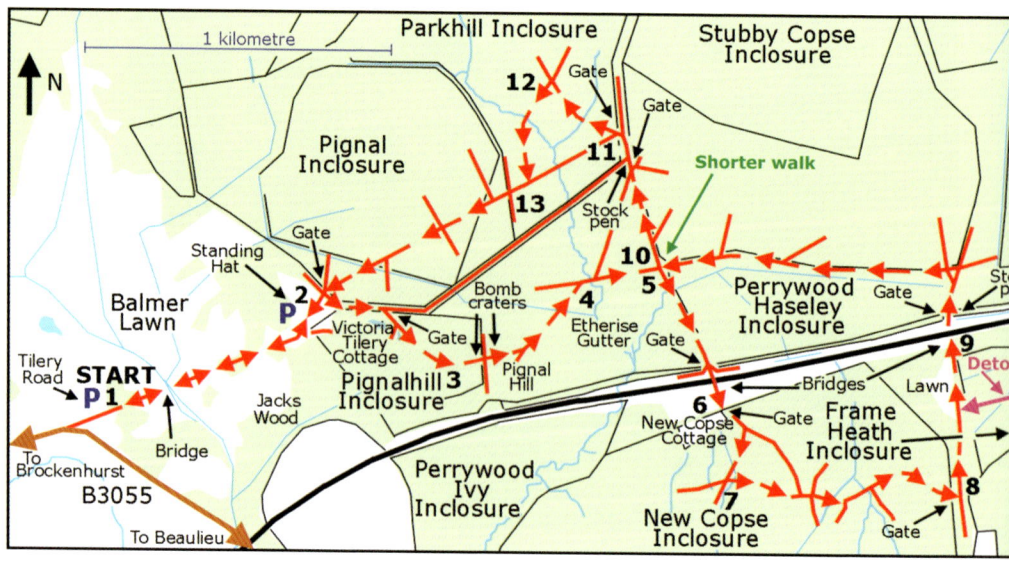

From Tilery Road car park

1. Turn left out of the Tilery Road car park; pass a low, Forestry Commission vehicle barrier and continue along a quite wide, gravel cycle track.

   Follow this over a railed bridge; pass on the left and right, grasslands dotted with clumps of gorse and broadleaved trees; and eventually go left at the first fork in the track, following directions on a sign advising: 'All Cyclists, Walkers and Vehicles this way' – straight ahead provides access only to private residences.

   Standing Hat, with its own small car park, is a short distance along here on the left, adjacent to extensive woodlands.

2. Immediately before entering the woodlands, turn right, opposite the Standing Hat car park entrance. Pass a low, Forestry Commission vehicle barrier and follow the gravelled cycle track as it eventually passes behind a cluster of buildings associated with Victoria Tilery Cottage.

   Continue along the cycle track as it bends to the right, ignoring here a wide, grassy driftway going straight ahead. Immediately pass through a gate and then almost immediately ignore a gravel track on the right leading to the Victoria Tilery Cottage buildings.

3. Follow the cycle track as it eventually bears left; and after a short distance, go straight ahead at a crossroads, with to the right, another gravel track, and to the left a grassy ride.

   A series of fairly shallow, Second World War bomb craters can be found a short distance to the left of the walk route, on either side of this crossroads.

> **Blackthorn – blossom to dazzle the eyes**
> Blackthorn blossom erupts in snowy waves some considerable time before the leaves appear, offering the mouth-watering prospect of abundant late summer and autumnal sloes. Particularly harsh weather during the blossom period gave rise to the traditional country term for a cold spring: a blackthorn winter.

77

Exactly when the bombs were dropped, and the identity of their originally intended target, is unclear.

After a further short distance, where the cycle track again bears left, pass a wide, grassy ride on the right.

4. Follow the cycle track as it eventually goes quite sharply right, ignoring here a grassy ride on the left and soon after, a second ride on the left.

5. Reach a crossroads of gravel cycle tracks.

    To take the shorter walk, which misses out the entire south-eastern part of the route, turn left here and follow the route instructions from Section 10.

    Otherwise, turn right, eventually leave the inclosure through a gate and continue uphill to cross the railway line at a bridge.

6. Go through a gate beside New Copse Cottage and follow the gravel track leading into New Copse Inclosure. Pass a wide, grassy ride going almost straight ahead where the cycle track bears right; and turn left at the next crossroad of gravel cycle tracks.

Badgers – boars wander widely
Under the cover of darkness, mature boar badgers spend much time patrolling their territories during the main February to April mating season. Presence is signified using strategically located dung pits and deposits of scent. Temporary excursions are undertaken to visit neighbouring badger groups, probably in search of mating opportunities. Residents are likely to show aggression to interlopers, and fights frequently occur. All too often, wanderers fall victim to road traffic accidents.

More...
Badgers – autumnal diet: page 158
Badgers – rarely seen, but signs betray presence: page 205

7. Follow the undulating track as it meanders through the woods, passing along the way, two quite wide rides on the left and three on the right. Eventually go through a gate and turn left at a 'T' junction of gravel tracks.

8. After around 200 metres, notice on the left, an area of close-cropped lawn bordered on its far side by unenclosed woodland; and reach on the right, a wide, gravel track leading into Frame Heath Inclosure.

To extend the route and increase the prospect of seeing sika deer, take a detour along the track running through Frame Heath Inclosure, which is a favourite haunt of these attractive creatures.

Otherwise, continue straight ahead, past a low, Forestry Commission vehicle barrier.

9. Follow the gravel track as it goes over a bridge across the railway line.

> **Wild flowers – the first of the year**
> In sheltered places free from excessive grazing, lesser celandine and primrose flowers – the former, bold golden-yellow; the latter, also yellow, but of a paler hue – make memorable first appearances during March.
> Common dog-violets at this time also produce their own dainty blue-violet blooms. Used as a derogatory term, 'dog' in this context differentiates the plant from the supposedly superior, sweet violet, but the name is scarcely warranted for in the New Forest, the common dog-violet is the most widely used caterpillar foodplant of pearl-bordered, small pearl-bordered and silver-washed fritillary butterflies, and is consequently of great value.
>
>
> *Primroses*

Adjacent to the bridge, on the left-hand side, a gate gives access to the track, and by the gate is a Network Rail sign that reads: 'Be aware that this is an access to a legally protected environment site'. The sign, however, is intended to notify railway workers of the special status of the New Forest, and does not draw attention to a special, track-side wildlife reserve.

Almost immediately pass on the right, a wooden stock pen used during the annual pony drifts. Go through a gate and after a short distance, turn left at a crossroads of gravel tracks.

Continue for 1 kilometre (0.6 miles) – ignore minor paths to left and right

> **Chiffchaffs – early visitors, simple song**
> Chiffchaffs, often the first spring and summer visiting birds to be noticed, usually start to arrive in the New Forest in late March, many after travelling from distant wintering grounds, others after spending the colder months more locally. Males immediately add song – a simple repetition of their name: chiffchaff, chiffchaff, chiffchaff – to the chorus of resident voices and will continue to do so until late June or July.

along the way – and then turn right at the crossroad of gravel cycle tracks passed during Section 5 on the outward route.

10. After a short distance, ignore a woodland ride on the right and after a further 200 metres, go straight ahead at a junction of grassy woodland rides. Pass through an adjacent gate and immediately after, pass a wide, grassy driftway on the left – there is another, smaller, stock pen here, too.

11. Continue straight ahead for a short distance before following the cycle track round to the left, through a second gate.
    Go straight ahead, again for a relatively short distance; follow the track as it bears right and turn left at the next crossroads.

12. Continue along the cycle track as it bears first to the left and then sharply right, past on the sharp bend, a grassy ride on the left.

13. Continue straight ahead through a corner of Pignal Inclosure, past woodland rides to left and right. Go through a gate adjacent to the Standing Hat car park and retrace the outward route back to the Tilery Road car park (and if required, to Brockenhurst railway station and the village).

For the adventurous, for those with a good sense of direction, strong map reading skills and access to an Ordnance Survey map!
Create your own walk by combining parts of this route with elements of your choice from the following selection of connecting or conveniently located nearby routes.

From *New Forest Walks – a seasonal wildlife guide:*
    April   Beaulieu Road: Shatterford
    September   Lyndhurst: Clay Hill

From *New Forest Walks – a time traveller's guide:*
    Walk 11   Brockenhurst: Balmer Lawn

Start	Shatterford, Forestry Commission car park, 5.5 kilometres (3½ miles) south-east of Lyndhurst on the B3056 Lyndhurst to Beaulieu road – Ordnance Survey map reference SU348064
Distance	9 kilometres (5½ miles) Shorter walk options: 1) Reduce the distance by 2.5 km (1½ miles) 2) Reduce the distance by 1 km (0.6 miles) Walk extension: A detour into Frame Heath Inclosure provides the potential to extend the route
Time to allow	2¼-5½ hours
Refreshments	The Drift Inn, near Beaulieu Road station, is close to the start of the walk
Route	Primarily along readily visible tracks
Terrain	Mainly level ground
Rating	2 – moderate walking
Buggies	Not suitable
Railway station	Beaulieu Road, adjacent to the start of the walk
Bus service	None
New Forest Tour Bus	Yes
Alternative starts	1) Beaulieu Road, Forestry Commission car park at Ordnance Survey map reference SU351063 2) Pig Bush, Forestry Commission car park at Ordnance Survey map reference SU362051 3) Yew Tree Heath, Forestry Commission car park at Ordnance Survey map reference SU365064
'Camping in the Forest' Caravan and Campsites	1) Matley Wood, 2.5 kilometres (1½ miles) 2) Denny Wood, 2 kilometres (1¼ miles)

APRIL
Wild flowers, butterflies and birdsong
(Beaulieu Road: Shatterford, Woodfidley, Rowbarrow, Pig Bush, Ferny Crofts, Yew Tree Heath and Black Down)

Bluebells in New Forest woodland

The walk
This heathland, woodland and wetland walk passes through one of the New Forest's best-known wildlife watching sites – the area around Shatterford Bottom, Rowbarrow and Pig Bush – before continuing over equally attractive, but perhaps lesser known countryside near Ferny Crofts and Yew Tree Heath.

Birds, wild flowers, butterflies, dragonflies and damselflies may all prominently feature. The 9 kilometre (5½ mile) route, mostly on level ground, can be reduced by up to 2.5 kilometres (1½ miles), whilst a detour into Frame Heath Inclosure offers the prospect of encountering sika deer.

Featured wildlife

Dartford warblers – tiny heathland denizens

If one bird epitomises the New Forest heathlands, it is the Dartford warbler. Tiny, long-tailed creatures no more than 14 centimetres (5½") in length, males are a little more boldly marked than females, although both sexes have wine-red under-parts and considerable amounts of brown and grey above.

Normally shy and secretive, these habitual skulkers in the undergrowth can be incredibly difficult to see, their presence betrayed only by occasional low level flights and harsh, buzzing calls made from the sheltered confines of heather or gorse. But in spring, and occasionally at other times, their scratchy, warbling song may be heard and dancing display flight witnessed.

Unlike many other warblers found in Britain, Dartford warblers do not migrate south for winter, although newly fledged youngsters may wander widely whilst dispersing. Instead, the majority stay at home,

Dartford warblers: a male and female obligingly perch in full view

adopting a somewhat risky strategy that results in wildly fluctuating population levels that are strongly influenced by the severity of winter weather.

When numbers are high, these attractive birds can be found on many New Forest heathlands, providing that heather tall enough to offer well-concealed nest sites is present, and gorse is available as living space for spider and insect prey items.

Dragonflies and damselflies – colourful, primarily wetland inhabitants

As spring gathers pace, insects increasingly take to the wing, including a sometimes bewildering range of dragonflies and damselflies. Although typically seen from late April until early November, flight periods vary from species to species. The first to be noticed will often be large red damselflies, followed in early May by broad-bodied chasers. Late fliers, those often on the wing until mid-autumn, include common darters, common hawkers, migrant hawkers and ruddy darters.

All need reasonably warm, dry, calm weather conditions in which to become fully active – few will fly in cold, wet or windy weather. Individual winged insects can live for two to three months, but for many, this phase of life is considerably shorter. In Britain's relatively cool climate, none survive the winter as adults.

A large red damselfly with insect prey

In their early life stages, all are aquatic, so watery habitats are often the best places to see the greatest number of species. However, some wander widely onto heathland, whilst others are frequently seen hunting insect prey along woodland rides.

More...
Southern damselflies – jewels in the New Forest crown: page 117
The Crockford Stream: page 118
Beautiful demoiselles – incongruously exotic insects: page 137
Deadman Bottom: page 146
Dragonflies and damselflies – late fliers: page 185

Hobby – a magnificent falcon

Hobbies, dashing spring and summer visiting falcons, are drawn each year to favoured New Forest heathlands and wetlands where they engage in impressively agile, yet graceful hunting flights as thy seek out large insects and small birds. The first of the year are usually seen at the beginning of April, although it is not until later in April that numbers significantly increase as new arrivals flood in from southern African wintering grounds.

In late May or early June, two or three eggs are laid in a sometimes regularly used nest, usually originally built by carrion crows in a woodland edge or heathland conifer. Growing chicks are often fed almost entirely on small birds

85

caught at height and brought back by parents that at this time increasingly enjoy the same diet. By late July or early August, fledged but still dependent youngsters, sometimes clearly visible perched in the upper, outer branches of coniferous trees, use loud, sharp, far-carrying calls to declare presence, alerting over-worked parents to the whereabouts of hungry mouths to feed.

Most hobbies leave the New Forest by late September at the start of their long migratory journeys, although small numbers sometimes linger into October.

Hobbies are dashing spring and summer visitors

Silver birch – pollen-laden herald of the spring

Pale barked, ever-graceful and once aptly known as 'ladies of the woods', silver birch trees are common and widespread throughout the New Forest. A relatively short-lived species – one of the largest recorded locally, with a girth of a little over 3 metres (118 inches), is thought to be less than one hundred years old – silver birch grows well in the New Forest's often acidic soils. Frequently the first trees to colonise new areas of clear-felled woodland, they can also be found around woodland edges where they extend the size of the wood as new ground is taken in. Browsing pressure and periodic, controlled burning, however, help to control excessive encroachment onto heathland.

A birch tree in spring, with witches' brooms clearly visible

In early April, the appearance of fresh, new birch leaves signifies the progress of spring just as effectively as anything else in the natural world, whilst at around the same time, male birch catkins swell, their bracts open and copious amounts of pollen are released to be carried on the breeze to the considerably smaller females. (Hay fever suffers may be dismayed to learn that a single catkin can contain five and a half million pollen grains.)

Conspicuously present amongst the pendulous branches, plant-galls – tangles of twigs sometimes known as witches' brooms – can be mistaken for birds' nests, but are, in fact, naturally formed growth deformities caused by the presence of a fungus.

Willow warblers and wood warblers – visitors from afar

Newly arrived from sub-Saharan Africa, willow warblers from early April sing their whispering, silvery cadence from vantage points high up in silver birch and other trees, and will continue to do so until June or early July, and again, occasionally, in late summer. Often widespread and abundant, these tiny,

Willow warblers produce a delightful, whispering song

primarily brown-green and pale yellow birds are almost identical in appearance to closely related chiffchaffs. Willow warbler song, however, is significantly different to that of the chiffchaff, a bird that is content to merely repeat its name.

Wood warblers are slightly larger than willow warblers, and are also spring and summer visitors from Africa. They occur in the New Forest in variable numbers from around the middle of April and add soft, piping notes and gentle trills to the cacophony of other bird sounds

A wood warbler beside its nest on the woodland floor

heard at dawn, dusk and intermittently throughout the day. Primarily yellow-green above and with a white belly, wood warblers prefer to live in fairly open, grazed woodlands, although close-planted, even coniferous, inclosures may sometimes be occupied.

Wood-sorrel – exquisite woodland blooms

Wood-sorrel flowers appear in early spring, usually before the trees come fully into leaf and cast all below in shade. Slender, red-tinged stems, rarely more than a few inches high, support a delicate, creamy-white, bell-shaped flower that boasts a series of thin, lilac or mauve veins running from the outer edge to deep inside the cup, lines evolved to help guide in tiny passing insects.

The veins are not always visible, however, for, to provide protection from cold and damp, the flowers close as light levels fall and open again only when the light improves; whilst in dull conditions the clover-like, heart-shaped leaflets fold downwards and droop beside the stem, thus also gaining protection from the elements. Similarly, when

Wood-sorrel plants produce exceptionally delicate blooms

exposed to excess sunlight, the leaves again fold down, preventing undue evaporation and overheating.

The spring-time flowers, though, bear little seed and are followed by more productive but far less noticeable self-pollinating blooms borne on shortened stalks.

Along the way
The area around Ferny Crofts and Yew Tree Heath has much to offer the wildlife enthusiast. Dry heath, wet heath and mires are all present, whilst the northern boundary of Yew Tree Heath is bordered by the Beaulieu River. Drainage channels crisscross the wetter places, small, permanent pools provide homes for aquatic creatures and ancient, unenclosed woodland is also represented.

Substantial gorse brakes and large expanses of heather attract Dartford warblers and stonechats all-the-year-round; linnets, meadow pipits, skylarks, greenfinches and nightjars are regular breeders; and great grey shrikes and hen harriers are autumn and winter visitors. Breeding waders are also present: displaying curlews can often be seen, lapwings nest in the drier places, and snipe and occasional redshanks inhabit the wetter ground.

A small pool on Yew Tree Heath

Dragonflies and damselflies hawk around the drainage channels and small ponds – look out, in particular, throughout the summer months for nationally scarce, small red damselflies and for the considerably more noticeable keeled skimmers – whilst marsh gentians are a delightful feature of late summer.

And, of course, the ancient woodlands (around Ferny Crofts and Decoy Pond Farm) will rarely disappoint those searching for wildlife.

The Route

1. Leave the Shatterford car park on its southern side – the side opposite the entrance; pass beside a low, Forestry Commission vehicle barrier and continue straight ahead along a quite wide, gravel track running broadly parallel to the railway line.

   Eventually cross a short length of causeway with railed bridge carrying the track across a very wet area, and then pass a track on the left leading over a railway bridge – this bridge offers an excellent vantage point for viewing the surrounding low-lying land.

   After a further 300 metres, follow the now sandy/gravel track through a gap in Bishop's Dyke, a low earthen bank and accompanying ditch that date back to the early 13th century – for part of the year, this is largely lost to view under bracken and other vegetation.

2. Continue straight ahead through an area of bracken, heading in the direction of Woodfidley Passage and Denny Lodge Inclosure. Cross another causeway, this carrying the track over a well-vegetated wetland area with here, two railed bridges in quick succession.

3. After a short distance, take the left-hand fork at a woodland edge 'Y' junction and follow the quite wide, grassy track round to the left and into the woodland.

> **Wild flowers – woodland opportunists**
>
> Bluebells, lesser celandines, wood anemones and wood-sorrel increasingly illuminate the woods, whilst bugle, germander speedwell, herb-robert, greater stitchwort, ramsons and wild strawberry blooms also make appearances. Grazing pressure from deer and commoners' stock restricts abundance and distribution, but some flowers usually survive to contribute much anticipated splashes of colour beneath the trees.
>
>
> *Lesser celandines*

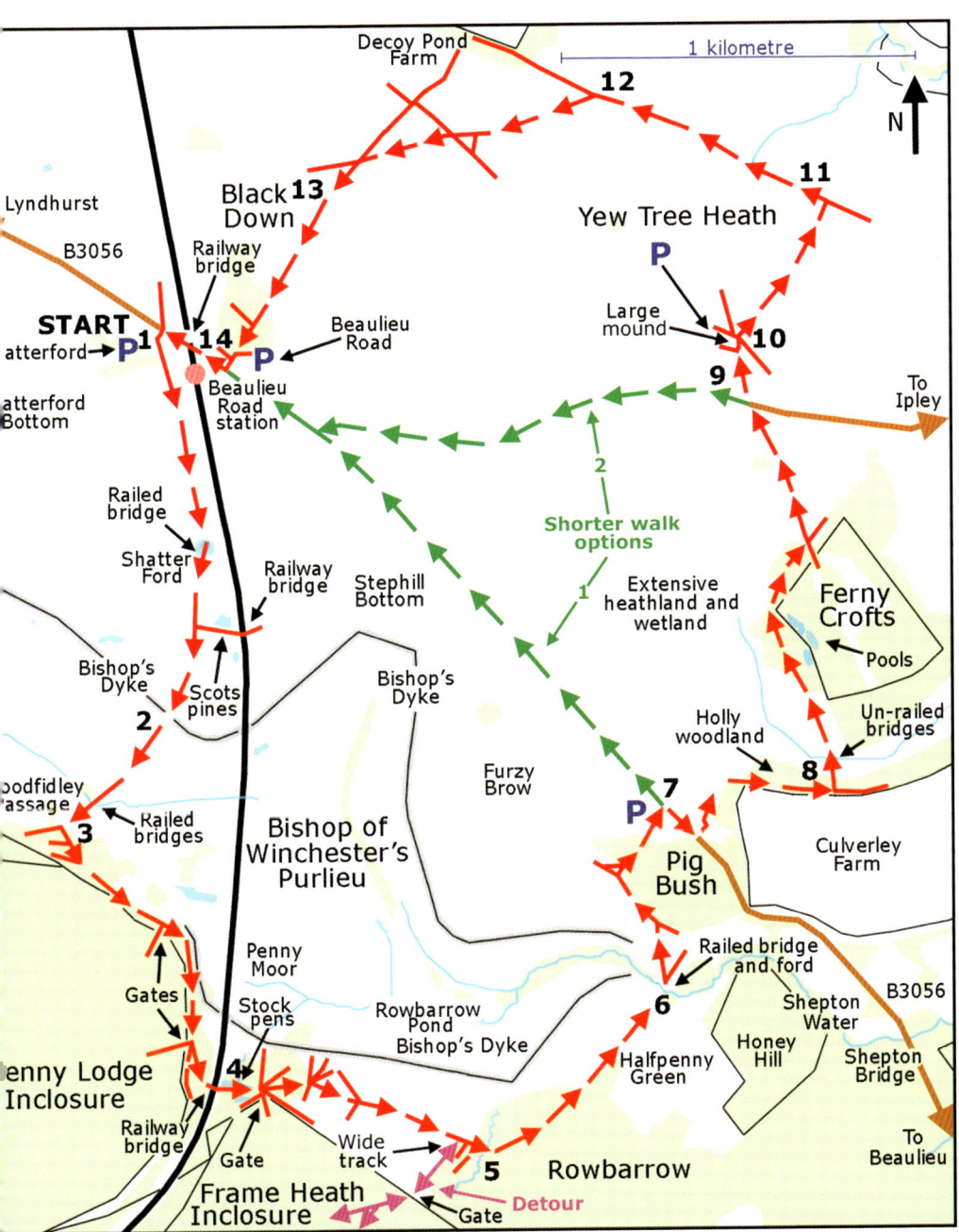

Ignore a narrow path on the right and soon after, a wider path on the right, before continuing close to, and parallel with, the Denny Lodge Inclosure wood-bank and fence, also on the right.

Eventually pass on the right, a gate leading into the inclosure; go up a short, gradual incline and follow the path as it bears round to the right. Continue along the path, parallel to the inclosure wood-bank and fence on the right; and eventually beside, on the left, another section of the Bishop's Dyke boundary bank.

Pass two gates together on the right, gates that from late spring to early autumn are usually shrouded in bracken; follow the track as it bears round to the left and cross the railway line at a bridge.

4. Follow the wide, largely gravelled track down from the bridge; pass through the gates of a stock holding pen and at the bottom of the slope, pass on the left, another stock pen.

   Immediately continue straight ahead at a junction of tracks to follow a wide, grassy track initially running through a small copse – at the junction, ignore a gate on the right leading into the inclosure, ignore tracks also on the right running alongside the inclosure and ignore two paths on the left, one often obscured by bracken.

   After a short distance, leave the woodland and continue along the main, quite wide, grassy track as it bears slightly right across open heathland. Ignore along the way, minor paths to left and right; and after around 450 metres, reach a quite wide track on the right, running alongside (on its left) an area of well-spread, encroaching trees (that include birches, hawthorns and crab apples).

   To extend the route and increase the prospect of seeing sika deer, turn right here and after around 200 metres, go through a gate that provides access to Frame Heath Inclosure, a favourite haunt of these animals. Walk along the gravel, inclosure tracks for as far as time and inclination dictate, and then retrace your steps.

   Otherwise, continue straight ahead and after a short distance, enter an extensive area of ancient, unenclosed woodland. Almost immediately pass a minor and then a wider woodland track coming in from the right, and cross a small stream – this is shallow and narrow at most times of the year, and in summer is often dry.

5. Follow the main path as it bears half-left, ignore minor paths to left and

92

right, and eventually leave the woodland behind as, first on the left and then also on the right, open heathland is reached.

6. Cross a narrow causeway and railed bridge – beside a ford – over Shepton Water; almost immediately turn half-left, off the main track, and continue along a heathland path leading towards ancient, unenclosed woodlands at Pig Bush.

Go left at an inverted 'Y' junction beside the wood and follow the track alongside the woodland edge. After a relatively short distance – just before the grey, sandy dirt track gives way to close-cropped grassland – turn right and follow a narrow, slightly uphill, woodland path. After a further short distance, follow the path as it bears to the right and is joined by a minor path on the left.

Continue along a somewhat wider, uphill track; reach the Pig Bush car park and leave by its main entrance.

> **Grey wagtails – colourful, primarily spring and summer visitors**
> The name might conjure images of dull coloured, nondescript birds, but that would be wrong for whilst parts of the grey wagtail are predictably grey, the breast and rump are often strikingly yellow, the flanks and belly are sometimes noticeably paler and much of the wings and tail are black. Nor are their habits grey, for from March until late summer or early autumn, grey wagtails are conspicuously present visitors to New Forest streams where they build well-concealed nests within, for example, crevices in stream banks and amongst tree roots close to water.
>
> Flight is invariably undulating and, inevitably, the bird's tail constantly wags up and down.
>
>

7. **To take the first shorter walk option**, which misses out the northern part of the route, turn left beside the minor road – the B3056 – and rejoin the main route at Section 14.

Otherwise, cross the minor road; turn right, along the grassy roadside verge and when around 40 metres from woodland up ahead, go left along a quite wide heathland track.

Continue half-to-the-left, alongside the woodland edge, before almost immediately following the path downhill, eventually to the right. Enter a corridor of predominantly holly woodland alongside on the right a

93

> **Breeding waders – snipe, curlews, lapwings and redshanks**
> Resident snipe and recent breeding season arrivals, at this time actively communicate using 'chipper' calls and drumming displays. Other visiting waders, meanwhile, renew spring and summer acquaintance with the New Forest's bogs, mires, wetter heaths and grasslands. Curlews announce presence with far-carrying calls, bubbling trills and conspicuous display flights; lapwings tumble through the air in their own erratic, elaborate displays and loudly call their alternative peewit name; whilst redshanks, 'yelpers of the marsh', birds that are present in the New Forest in seemingly ever decreasing numbers, react with anguished cries to suspected, often distant, sources of danger, invariably alerting other wetland inhabitants to the perceived threat.
>
>
> *Lapwing*
>
> More....
> Snipe – inhabitants of swamps and other wet ground: page 59
> Curlews – unmistakable wetland waders: page 99

substantial earthen bank with beyond, the grasslands of Culverley Farm. Eventually go up a short incline and as the ground starts to drop downhill, turn left along a narrower path.

8. Almost immediately cross a narrow, valley-bottom bog using un-railed bridges in quick succession, and then follow the path as it bears left, uphill, running parallel to the edge of broadleaved woodland away to the right.
   Enter a corridor of ancient, unenclosed woodland; pass on the right, a series of small pools; and continue along the path as it runs broadly parallel to, on the right, the wood-bank and fence separating the open Forest from Ferny Crofts, a Hampshire Scouting camping, training and activity centre.
   Eventually follow the path as it bears right, still quite close to the Ferny Crofts boundary; after a relatively short distance, turn left along the gravel, Ferny Crofts approach road and after 400 metres, reach a minor road.

9. **To take the second shorter walk option**, which misses out much of the northern part of the route, go left, continue beside the minor road and

then turn right at the next minor road – the B3056 – to rejoin the main route at Section 14.

Otherwise, cross the minor road and continue straight ahead along the Yew Tree Heath car park access road. As this eventually goes left, continue straight ahead for a further very short distance, adjacent to one side of a quite large, earthen mound on the left.

Its man-made origin betrayed by protruding concrete, this mound was the site of a Second World War anti-aircraft battery command post. The position of the associated guns, brought into action to help protect Southampton from enemy bombers, is revealed by a series of lower mounds still visible in crescent pattern around the edge of the car park.

10. Immediately go half-right – in the direction of Marchwood Power Station's distant, twin, pale chimneys – along a quite wide, grassy track, ignoring here other minor heathland paths to left and right. After a short distance, follow the track downhill; cross an often wet patch of ground and eventually turn left at a pronounced 'T' junction.

11. Follow what is now a quite wide, sandy/gravel track as it heads in the direction of broadleaved woodland close to Decoy Pond Farm, 1 kilometre (0.6 miles) away.

12. When around 200 metres from this woodland, take the left-hand fork at a 'Y' junction – the right-hand fork here almost goes straight ahead. Follow the track uphill, and quite soon after the ground levels out and the track bears slightly right, pass on the left a narrower sandy/gravel track.

Wheatears – just passing through

Wheatears very occasionally breed in rabbit burrows on local heathlands. However, following their arrival from Africa, passage birds are quite often seen in spring as they rest awhile before moving on to raise families in northern England and elsewhere. Small numbers are also present in summer and autumn whilst undertaking the reverse journeys.

> **Spring butterflies – most welcome sights**
> Red admirals, primarily migratory insects from North Africa and continental Europe, and painted ladies, butterflies also from North Africa, can be seen on the wing in April; and so can resident brimstones, commas, holly blues, peacocks and small tortoiseshells.
> Speckled wood butterflies are also likely to be encountered. Merging well with the dabbled light and shade of favoured woodland habitats, these chocolate-brown, creamy-yellow blotched insects fly until early October, albeit with short, late spring and mid-summer 'between brood' absences. Males are particularly eye-catching as they engage in territorial display flights in patches of sunlight.

Red admiral

Almost immediately go straight ahead at a crossroads and after 250 metres, turn left along a tarmac surfaced road serving Decoy Pond Farm.

13. After a further 500 metres, follow the road as it enters a clump of Scots pines, pass a vehicle barrier and gravel track on the right, pass the Beaulieu Road car park on the left, and pass another vehicle barrier and gravel track on the right, this leading to the nearby pony sale pens.

    Almost immediately turn right along a minor road – the B3056 – opposite the Beaulieu Hotel and adjacent pub.

14. Follow beside the road as it goes over a railway bridge, and the Shatterford car park is immediately on the left.

For the adventurous, for those with a good sense of direction, strong map reading skills and access to an Ordnance Survey map!
Create your own walk by combining parts of this route with elements of your choice from the following selection of connecting or conveniently located nearby routes.

> From *New Forest Walks – a seasonal wildlife guide:*
> March   Brockenhurst: Balmer Lawn
>
> From *New Forest Walks – a time traveller's guide:*
> Walk 8   Beaulieu Road: Black Down
> Walk 9   Beaulieu Road: Shatterford

MAY
Spring in all its glory
(Burley: Turf Hill, Holmsley Bog, Castleman's Corkscrew, Holmsley Inclosure and Shappen Hill)

The walk route follows this track across Holmsley Bog

The walk

Starting at Burley car park, this 7 kilometre (4¼ mile) walk offers impressive views over extensive heathlands and the potential to see wetland wildlife in and around Holmsley Bog. Castleman's Corkscrew, a railway line abandoned long ago and now used by walkers, cyclists and others seeking recreation, is

Start	Burley, Forestry Commission car park, almost opposite the track alongside Burley Primary School on the minor road leading uphill from the Queens Head towards the village cricket pitch, and on to the A35 at Holmsley – Ordnance Survey map reference SU214028
Distance	7 kilometres (4¼ miles) Shorter walk options: 1) Reduce the distance by 3.5 km (2 miles) 2) Reduce the distance by 1.25 km (¾ mile) 3) Reduce the distance by 0.25 km (¹⁄₆ mile)
Time to allow	1¾ - 4¼ hours
Refreshments	The Queens Head and the Burley Inn are both in Burley village
Route	Primarily along readily visible tracks
Terrain	Some undulating ground, but with few significant gradients
Rating	2 – moderate walking
Buggies	Not suitable
Railway station	Sway, 9.5 kilometres (6 miles)
Bus service	More/Wilts and Dorset, primarily on Mondays, Wednesdays and Fridays only
New Forest Tour Bus	Yes
Alternative starts	1) Burley Cricket, Forestry Commission car park, just across the road from Burley car park at Ordnance Survey map reference SU215029 2) Public car park beside the Queens Head, close to Burley village centre, at Ordnance Survey map reference SU211031 3) Holmsley, Forestry Commission car park at Ordnance Survey map reference SU222011
'Camping in the Forest' Caravan and Campsites	1) Holmsley, 5.5 kilometres (3½ miles) 2) Setthorns, 7.75 kilometres (4¾ miles)

followed before the route plunges into woodland. Return is via the old railway line and more, extensive wetland and heathland. Short-cuts are available to reduce the distance by up to 3.5 kilometres (2 miles).

Featured wildlife

Curlews – unmistakable wetland waders

Quite large, long-legged birds with lengthy, decurved bills, curlews are welcome breeding season additions to many of the more extensive New Forest bogs and mires. The first new arrivals appear from mid- to late March. Males immediately start to mark out their territories using conspicuous display flights that feature steep, upward climbs on rapidly beating wings followed by effortless glides as height is progressively lost. A variety of evocative, far-carrying calls accompany the performance: slow, drawn out *ooorr-ooorr-ooorr* notes, noisy repetitions of *curlee-curlee-curlee* and mellow, bubbling sounds.

Nests, simple depressions lined with vegetation and maybe a few feathers, are sited on the ground, often on a tussock or low hummock, and almost invariably in places relatively inaccessible to man. Nervous birds largely unfamiliar with the concept of sitting tight and letting danger pass, curlews are likely to take flight at the slightest hint of provocation and will circle round, noisily calling with ever-increasing signs of agitation until the threat has passed. Unintentional disturbance by people and dogs, for example, is therefore a significant problem.

Curlews attract attention with conspicuous display flights

Virtually all curlews leave the New Forest by July or August, many to spend the autumn and winter on the coasts of southern and south-west England, south Wales and southern Ireland.

More...
Breeding waders – snipe, curlews, lapwings and redshanks: page 94

Foxes – parents under pressure whilst raising cubs
Fox cubs, typically four or five, sometimes six, to a litter, are often born in March and initially remain underground in dens that are frequently located in commandeered sections of badgers' setts or in rabbit burrows specially enlarged for the purpose. Only from mid- to late April do they regularly emerge, usually at dusk but sometimes also by day, to explore the immediate area around the den and to play, loaf and patiently await food deliveries from parents kept ever-busy by the hungry youngsters.

Rabbits, other small mammals, birds as large as pheasants and virtually anything else edible that can be carried will be brought back, deposited on the ground and left for the family to squabble over; or else will be taken from the returned hunter's jaws by a particularly eager youngster. In times of

Fox cubs at play

plenty, all feed well, but when insufficient solid food is available, some of the cubs may continue to suckle until around twelve weeks old.

Den occupation, particularly in the later stages of residence, is frequently betrayed by the scattered presence of cub droppings and leftover foodstuffs, items such as rabbits' legs, squirrels' tails and birds' wings and feathers.

For a number of weeks after first emerging from the den, cubs often show relatively little fear of people and can sometimes be watched from quite close range, although this refreshingly naïve innocence of youth gradually gives way to more sensible circumspection. By mid- to late June, well-grown youngsters can sometimes be seen at dusk, wandering quite far from home. Sadly, many are killed on the roads at this time and also as they disperse a little later in the year.

More....
Foxes – opportunist feeders: page 187
Foxes – loud cries disturb otherwise quiet winter nights: page 206

Pearl-bordered and small pearl-bordered fritillaries – rarities, both

Once common and widespread in much of the English countryside, pearl-bordered fritillaries have disappeared from many areas, victims of changed woodland management practices such as the large-scale cessation of rotational hazel coppicing that previously provided a succession of ideal, inter-linked habitats that included quite large areas of potentially sunny, open ground in which wild flowers – larval foodplants (most often common dog-violets) and sources of nectar for the adults – could flourish. In the New Forest, however, these insects remain notably present, although distribution is restricted.

Pearl-bordered fritillary

Relatively small, golden-orange coloured creatures with on the wings, a network of fine black veins, crossbars and spots; pearl-bordered fritillaries on the underside of the hindwings also show two prominent silver 'pearls' and the row of seven outer 'pearls' that give them their name. In flight mainly from mid-May until mid-June, they are most active in bright light – when the sun goes behind a cloud they almost immediately seek shelter, often coming to rest amongst woodland grasses.

Small pearl-bordered fritillaries fly a little later – typically from late May to early July. Quite similar in appearance to their virtually same-sized cousins, they are very scarce in the New Forest, and in most places elsewhere, but do occur in Holmsley Inclosure, visited during this walk.

Small pearl-bordered fritillary

Red deer – huge, majestic beasts

Britain's largest wild, land mammal, red deer in the New Forest are most often seen on and around a small number of favoured heathlands and in nearby woodlands. Current Forestry Commission policy is to maintain numbers by selective culling at a maximum overall spring population level of around 125 animals, whereas in past centuries, even though the red deer is a native species, periodic re-introductions – the most recent in 1962 – have been necessary to maintain a viable population.

Summer coats, acquired in spring, are an unspotted reddish-brown colour; winter coats, acquired in autumn and also unspotted, are dark brownish-grey. Rumps are usually paler than the body colour, whilst the relatively short tail almost matches the main coat colour. Prior to the autumnal rut, stags develop a noticeably thick, dark mane.

As with the other species of New Forest deer, red deer antlers are worn only by the males, and on mature animals are typically long, round in section and multi-pointed – twelve points (six on each side) are sometimes seen, whilst small numbers of particularly well-endowed stags boast even more. Yearling stags' antlers are simple, relatively straight, un-branched spikes.

Old antlers are cast in March, re-growth begins almost immediately and they are free of 'velvet' before the mid- to late September start of the annual rut when, for three to four weeks, the heaths, particularly early and late in the day, echo to the sound of roaring stags.

A huge red deer stag, its coat muddied during the rut

Births, usually of a single calf, take place in June. Hinds – the females – and youngsters can regularly be seen throughout the year, but outside the rutting season, many well-grown stags wander widely, often onto farmland or into woodland beyond the boundary of the Crown Lands.

Reptiles – far better behaved than their reputation suggests

Although relatively infrequently seen, Britain's native reptiles – adders, grass snakes, smooth snakes, common lizards, sand lizards and slow worms – probably occur in greater numbers in the New Forest than in most other places of comparable size.

Grass snakes and slow worms often tend to be associated with gardens and similar places close to human habitation – grass snakes like to feed on fish and amphibians, and therefore have a particular liking for ponds (including garden ponds), whilst slow worms are a little more catholic in their choice of living arrangements and can sometimes be found in open woodland and on heathland. The other species are primarily heathland dwellers although some may also occasionally be seen in woodland clearings and along open rides, providing that generous amounts of ground cover are present.

Cold-blooded creatures that regulate their body temperature by taking in, or avoiding, warmth from external sources, reptiles often bask in early morning sunshine, sometimes on exposed ground and, in some cases, within the upper sections of heather, soaking up the rays before going about their daily business. (Watch out especially for momentary glimpses of rapid movement as common lizards dart from sight to take cover amongst heathland vegetation, and for adders basking on south-facing banks.)

All spend late autumn and winter in hibernation, sometimes communally, tucked away in a burrow originally belonging to a rabbit, rodent or other animal; or concealed amongst a log pile, hidden under tree roots or anywhere else that

103

offers cool, but not too cold, temperatures, shelter and safety.

Only the adder is poisonous, yet even they prefer to avoid confrontation with humans and will not strike unless deliberately or accidently antagonised, but instead prefer to slink into cover until any perceived threat has passed. (Fatalities are extremely rare – there have been only fourteen, or so, recorded cases in the last hundred years.)

A slow worm crosses a track near Eyeworth Pond

More....
Reptiles – late summer youngsters: page 145

Stonechats – bold, brash and noisily conspicuous

Dumpy, robin-like birds, stonechats are the archetypal heathland sentinel, creatures that when confronted by even the slightest hint of threat, draw attention to themselves with harsh, scolding *tsak-tsak-tsak* calls given from atop a sprig of gorse or from a perch in the lower branches of an invading birch or other tree. (The calls are similar in sound to the noise made when two stones are knocked together, which explains the derivation of the birds' name.)

Males have reddish chests and bold, black heads; females possess duller, but similarly patterned plumage. For much of the year, stonechats are usually relatively abundant on New Forest heathlands, but they are, however, partial migrants – from late summer onwards, some travel south into mainland Europe and occasion-

Stonechats habitually perch in full view atop sprigs of gorse

ally beyond, whilst others remain on the breeding grounds unless forced to move by the onset of particularly cold weather.

Along the way

Crisscrossed by myriad minor tracks, the extensive heathlands south of Burley village host a range of wildlife species unique to places such as this, whilst, although perhaps less aesthetically pleasing, Holmsley Bog also warrants attention, particularly in spring and summer when it is home to an array of wild flowers, dragonflies, damselflies and breeding birds.

Walks here in autumn and winter also often provide wildlife interest, for hen harriers may be seen hunting over both heath and bog, merlins and peregrine falcons are sometimes glimpsed and a great grey shrike may also be present, patiently waiting on a favourite perch to pounce on passing prey items.

A minor track on Turf Hill

The Route

1. Leave the Burley car park from beside its main approach road, and go south-east through a line of three short, wooden posts that separate the car park from the adjacent heathland.

105

Immediately pass a gravel path on the right – this is used at the end of the walk. Follow the wider, more prominent gravel track, running at close to 90 degrees to the car park edge, as it passes through gorse on a course broadly parallel to the nearby minor road.

Almost immediately pass another lesser gravel track on the left; and opposite, a grassy path on the right. Subsequently ignore a number of minor paths to left and right, and continue across the heath along what quickly becomes a gently meandering, part-gravel surfaced track running for much of the way along a quite broad ridge, a plateau that gives way on both sides to lower ground – extensive views are available to left and right, but only if a detour over the plateau is made.

Pass on the left, a somewhat incongruously located golf course green and further along, also on the left, the gorse, heather and bracken-clad slopes of Broadoak Bottom, a shallow valley running roughly parallel to the track.

2. Eventually go steeply downhill to reach Holmsley Bog, a quite extensive wetland area rich in wildlife, and continue across a narrow, raised causeway. Almost immediately ignore a path going uphill on the left and another on the right; and continue straight ahead, uphill along the main track.

After a short distance, go downhill to reach at a junction, the old trackbed of Castleman's Corkscrew and the crumbling brickwork remains of Greenberry Bridge.

3. **To miss out the entire south-eastern part of the route**, take the first shortcut by turning right at this 'T' junction and following the walk directions from Section 13.

Scots pine – flowers that often go unnoticed

Scots pine flowers, all too often overlooked amid the melee of other spring-time activity, gleam in the sunlight. Indeed, particularly conspicuous clusters of large, pale male flowers can easily be mistaken for fresh, new cones. Carried on the breeze, clouds of golden pollen, dust the far less noticeable female flowers that will eventually go on to produce the new cones.

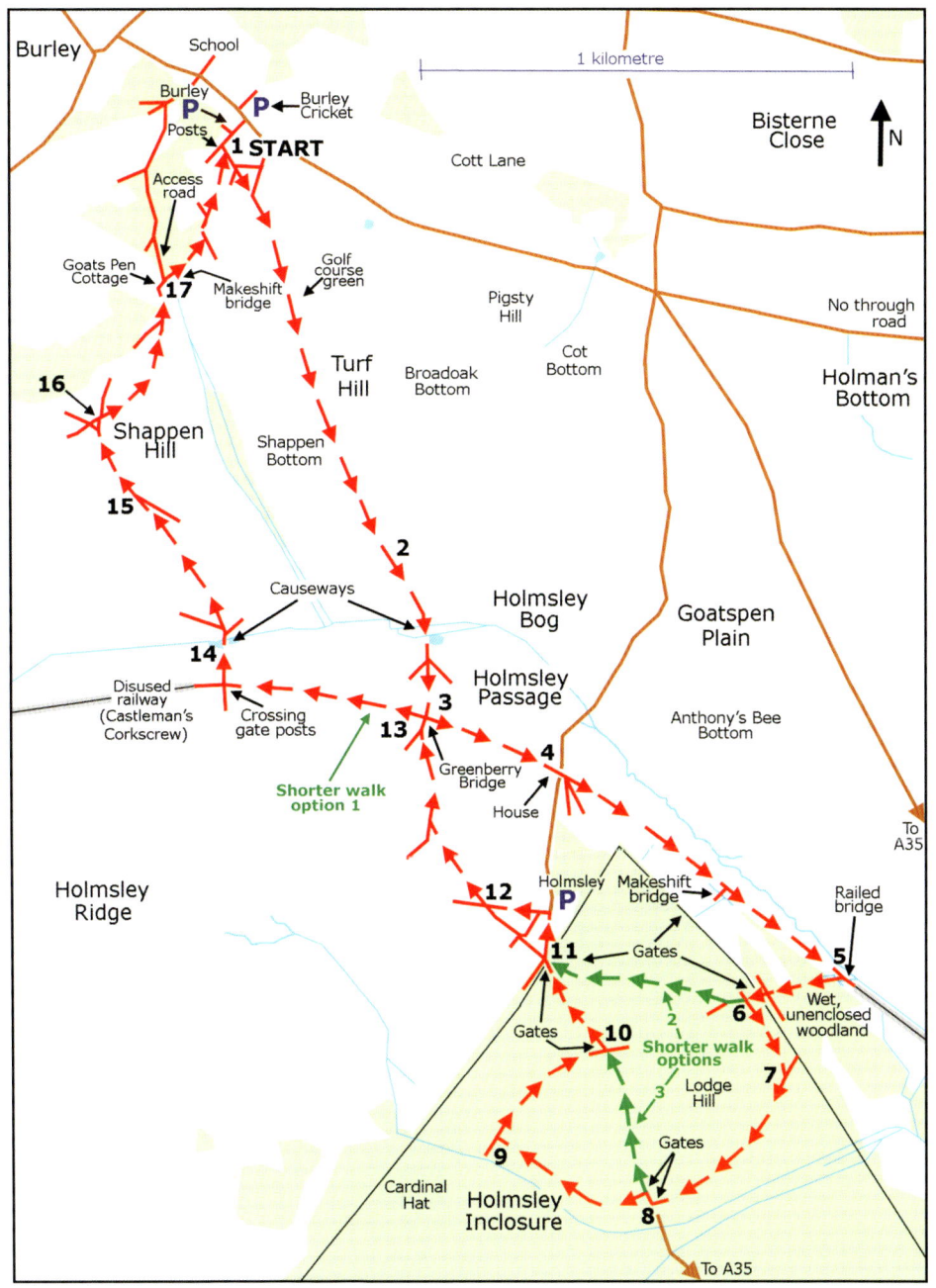

Otherwise, turn left and go along this quite wide trackway as it initially passes through a cutting.

4. Eventually pass a relatively modern house on the right, go beside a Forestry Commission vehicle barrier, cross a minor road, pass another barrier and continue along the trackway.

   Pass on the right, a makeshift bridge – a single railway sleeper, or similar – over a ditch, with beyond, an indistinct path leading across a narrow, grassy heath towards a pedestrian gate that gives access to Holmsley Inclosure. Immediately after this makeshift bridge, enter a tree-lined section of the trackway.

5. Eventually reach a substantial, metal-railed bridge that carries the trackway over a quite wide channel taking water from adjacent wet, unenclosed woodland into Holmsley Bog.

   Turn right immediately before, and adjacent to, the bridge; follow the path down the embankment; cross a low, plank bridge over a narrow stream and continue along a gravel track raised slightly above the level of the surrounding wet, primarily alder woodland.

   Continue straight ahead through a pedestrian gate and into Holmsley Inclosure.

6. **To take the second shorter route**, which avoids much of Holmsley Inclosure, almost immediately continue straight ahead at a crossroads and shortly after, bear slightly right, uphill, at a 'Y' junction. Eventually go through a gate adjacent to a minor road, cross the road, turn right and follow the main route directions from mid-way through Section 10.

> **Down in the woods – a busy time for birds**
>
> Woodland birds of all sizes and descriptions, evolved to produce as many fledglings as possible in often implausibly short timescales, pack every ecological niche, from the woodland floor right the way up to the treetops, during the often frantic, May breeding season climax. Nest site locations are often betrayed by careless, calling youngsters keen to encourage parents to feed them, before their siblings.

Blue tit

> **Meadow pipits – obliging surrogate parents on behalf of absent cuckoos**
> Some of them recently returned from as far away as Spain and Portugal, in spring meadow pipits cast off winter reticence and engage in extravagant 'parachute' display flights during which they launch themselves upwards, often from the ground, although sometimes from a perch, and climb steeply skywards before plunging back to earth, wings bent stiffly back, singing as they go.
> Cuckoos, grateful for the meadow pipits' presence, announce arrival with familiar calls and lay eggs in the nests of these unsuspecting hosts before, sometimes as early as June, embarking on return journeys to African wintering quarters. Many young cuckoos are thus abandoned, left behind to fend for themselves in an often hostile world and to make their first migratory journeys unaided. (Adult males typically depart before the females – it is thought that females store sperm, and continue laying eggs after the males have gone – whilst the youngsters of the year are usually the last to leave.)

Otherwise, turn left at the crossroads and continue along a quite wide, gravel track.

7. Eventually pass a grassy track on the left, follow the gravel track as it twice bears right and go through a pedestrian gate leading onto a minor road.

8. **To miss out the western part of Holmsley Inclosure**, take the third and final shorter route by turning right here and continuing alongside the road to rejoin the main route at Section 10.

   Otherwise, cross the minor road; turn right, and immediately left through another pedestrian gate; and follow a gravelled cycle track as it quite soon bears round to the right.

Nightjars – voracious enemies of moths and other night flying insects
Enigmatic, largely nocturnal birds newly arrived from African wintering quarters, nightjars in May take up residence on the heaths and in areas of clear-felled woodland. Characteristic churring song is most often heard at dusk, although disturbance by distant gunshots, rolls of thunder, passing trains or any other sudden noise can prompt short bursts of day-time sound.
Nightjars nest on the ground and rely on cryptically patterned plumage to avoid detection; and whilst trawling the night skies, skillfully employ exceptionally agile flight and a particularly wide gape to catch moths and other insects.

9. Turn right at the next 'T' junction to follow a fairly wide, moderately uphill, track before eventually, at the top of the hill, going through a gate and turning left beside the minor road.

10. Continue alongside the road; eventually pass on the right the gate used at the end of the second, shorter route (Section 6); and a pedestrian gate on the left. Immediately after leaving the woodland, pass a wide, gravel track on the left.

11. Continue alongside the minor road as it bears a little to the right, over a cattle grid; and after around 100 metres, turn left along the first wide, gravel track encountered. Almost immediately ignore a track on the left and pass beside a low, Forestry Commission vehicle barrier.

12. After a further 100 metres, turn right at a minor crossroads to follow along a heathland track that soon bears a little to the right and eventually goes downhill.
    Pass two narrow paths on the left and eventually turn left at Greenberry Bridge.

13. Follow the old railway trackbed for around 500 metres. Eventually, upon reaching four old gate posts – two pairs, where crossing gates once stood – turn right, downhill, along a gravel track.

14. Cross another narrow causeway over Holmsley Bog; at the end of the causeway, pass a minor track on the right and then a very indistinct path on the left; and follow the main track as it goes slightly left, uphill and on over the heath.

15. Eventually, at an inverted 'Y' junction, join a prominent gravel track. Follow this as it bears slightly left and take the next right-hand fork at a 'Y' junction not far from a block of woodland lying straight ahead.

16. After a very short distance, turn right at a 'T' junction; almost immediately pass a minor track on the left and continue downhill along a gravel track. Follow this as it gradually bears left; pass another minor track on the left; go beside a low, Forestry Commission vehicle barrier in front of Goats Pen Cottage and turn right, along the gravelled access road.

17. Almost immediately – when just beyond track-side garages – as the access road goes sharply left, continue straight ahead over a narrow strip of sometimes damp grassland and cross a small, often wet channel, if necessary using planks already placed to form a makeshift bridge.

    Follow a narrow, meandering path uphill, alongside woodland edge on the left. Eventually join a more substantial gravel track at an inverted 'Y' junction and go half-left, still uphill. Almost immediately pass a gravel path on the left and continue straight ahead to the Burley car park, which is a short distance away on the left.

    Note: After wet weather, the Goats Pen Cottage access road might provide a preferable alternative route back to the car park.

For the adventurous, for those with a good sense of direction, strong map reading skills and access to an Ordnance Survey map!
Create your own walk by combining parts of this route with elements of your choice from the following selection of connecting or conveniently located nearby routes.

    From *New Forest Walks – a time traveller's guide:*
    Walk 14   Burley: Shappen Hill

Start	Crockford, Forestry Commission car park, 5.5 kilometres (3½ miles) south-west of Beaulieu on the B3054 Beaulieu to Lymington road. Ordnance Survey map reference SZ350989
Distance	4 kilometres (2½ miles) Shorter walk option: Reduce the distance by 1.75 km (1 mile) Walk extensions: Detours to potential dragonfly and damselfly hotspots can add up to 2 km (1¼ miles) to the distance
Time to allow	1 - 2½ hours
Refreshments	Montagu Arms Hotel, Beaulieu; Turfcutters Arms, East Boldre
Route	Mostly along well-defined, readily visible tracks, but in places – for example, during Section 8 – a little 'off the beaten track'
Terrain	Mainly on level ground. (After heavy or prolonged rain, and in winter, the ground in places – for example, around the stream during Section 9 – can be rather wet. The use of tall, waterproof footwear, such as Wellington boots, is at these times recommended)
Rating	1 – easy walking
Buggies	Not suitable
Railway station	Lymington, 4 kilometres (2½ miles)
Bus service	More/Wilts and Dorset serve nearby Hatchet Pond, primarily on Tuesdays, Thursdays and Saturdays only
New Forest Tour Bus	Yes
Alternative starts	1) Norley Wood, Forestry Commission car park 1 kilometre (0.6 miles) south-west of Crockford car park, at Ordnance Survey map reference SZ346981 2) Crockford Clump, Forestry Commission car park at Ordnance Survey map reference SZ352992 3) Beaulieu Heath, Forestry Commission car park, near the Model Aircraft Flying Area at Ordnance Survey map reference SU358006 4) Hatchet Moor, Forestry Commission car park at Ordnance Survey map reference SU366011
'Camping in the Forest' Caravan and Campsites	1) Roundhill, 8.75 kilometres (5½ miles) 2) Hollands Wood, 11.25 kilometres (7 miles)

Note: In recent years, the Crockford and Crockford Clump car parks have been subject to seasonal site closures from March until July, closures intended to benefit the birds by reducing disturbance. Check the Forestry Commission website for the latest information or simply be prepared to use an alternative start point for the walk.

JUNE
Long summer days made for walking
(Crockford Bridge, Beaulieu Heath,
Crockford Stream and Shipton Holms)

A dog enjoys a cooling drink in the stream near Upper Crockford Bottom

The walk
The Crockford Stream, nearby ponds and areas of bog and mire are renowned for the presence of dragonflies and damselflies, whilst this walk also prominently features wide expanses of heathland that are home to a variety of special birds, wild flowers and insects. The relatively short, 4 kilometre (2½

mile) route, mostly over level ground, can be further reduced by 1.75 kilometres (1 mile) or extended to follow dragonfly and damselfly-related detours.

Featured wildlife

Bog myrtle – a source of wonderfully sweet aromas

A deciduous shrub of damp, open places, bog myrtle is common and widespread in and around New Forest wetlands where it grows to a height of up to 130 centimetres (51 inches). Small, resinous, aromatic glands on the leaves, twigs, stems and fruits exude a deliciously sweet fragrance that explains one of its popular alternative names: sweet gale – gale, until relatively recently in common use, was the early English name for the plant. However, Forest folk of old – from at least the early 19th century, and probably long before – knew bog myrtle as gold withy, and blamed it for giving a peculiar flavour to the milk of cows that ate it.

Catkin-like structures appear in April and May, usually before the leaves unfurl. Strikingly orange males brighten the wetland scene, whilst shorter, reddish females, usually found on separate plants, make their own subtle contribution to the landscape.

The scent, though, is not needed to attract insects, for these plants are wind pollinated. In fact, bog myrtle repels fleas and a variety of other insects, so-much-so that it used to be placed amongst linen to keep moths away and is still the main ingredient in a number of commercially available insect repellents.

John Gerard, herbalist and writer, in the late 16th century also noted that it gave a headiness to beer or

Bog myrtle 'catkins'

ale which was then 'fit to make a man quickly drunke' – gale beer was traditionally made from bog myrtle – whilst the aromatic resins have been frequently used for scenting candles.

Heath spotted orchids – delicate blooms illuminate the landscape

The New Forest is home to an impressive range of wild orchids, although many grow in relatively small numbers and are locally distributed. Heath spotted orchids, however, conspicuously brighten many of the drier heaths during the main late May to early July flowering period.

Heath spotted orchids are common and widespread

In fertile ground, they grow to a height of up to 40 centimetres (16 inches), although this height is not often achieved in the New Forest. Blooms, in a conical spike clustered around the tip of the stem, are most often pale pinkish-mauve, but white flowers and darker, particularly heavily marked examples also occur. Narrowly oval, pointed leaves, most prominent around the base of the plant, are usually marked with distinctive dark blotches.

Closely related common spotted orchids also occur and can cause identification confusion – subtle differences in each bloom's lip shape and markings help separate the two – but the presence of whichever happens to be encountered can be enjoyed without the need for absolutely definitive recognition.

Linnets – birds that enliven the heaths

Linnet numbers throughout much of Britain declined rapidly between the mid-1960s and the mid-1980s, probably as a result of changed agricultural practices, and have never since recovered, but these birds remain prominently present in the New Forest. Small, primarily brown finches, the males brightened with splashes of red, linnets do not usually over-winter locally, but in March and April return from as far away as southern Europe to add to the heathland scene twittering, conversational medleys given from favoured gorse bush perches or from amongst the branches of encroaching birch or pine trees.

115

Often semi-colonial breeders, linnets frequently build their nests, cosily lined with horse hair and soft feathers, in the midst of dense gorse bushes where, amongst the prickly foliage, they are relatively safe from predation. Their breeding season extends through to August.

Redstarts – colourful creatures of the woods

Known in days-gone-by as fire tails, the redstart's modern name derives from their conspicuous rust-red, constantly quivering tail – *steort* is Old English for tail. A relative of the robin and of similar size, redstarts are spring and summer visitors that arrive from Africa from late March and usually depart by September. In the New Forest, they nest primarily in natural tree cavities and, reflecting a strong preference for feeding on bare or sparsely vegetated ground, occur in many heavily grazed ancient, unenclosed woodlands, but are significantly less often present in woodland inclosures that are, or have relatively recently been, fenced to exclude commoners' stock.

Song, a hurried snatch of notes that peters out into a tinny jangle, given from high in the branches of a beech or oak tree, is often the first indication of presence, even though male redstarts are amongst the most colourful of British birds with plumage that includes a glorious mixture of orange-red, grey, black

Linnets frequent the heaths in spring, summer and early autumn

A male redstart: one of our most colourful woodland birds

and white. Females, however, are altogether duller coloured creatures: primarily grey-brown above and buff-white below.

Silver-studded blues – supreme heathland butterflies

On the wing from late June until mid-August, silver-studded blue butterflies are extremely scarce in much of Britain, yet on many New Forest heaths, particularly where the vegetation is kept relatively short by grazing or fire, these tiny, colourful, primarily sedentary insects are widespread and often abundant. Indeed, they are often the butterflies most likely to be seen.

Males are absolutely distinctive. Upperwings are deep blue with broad black margins and an outer fringe of white, whilst the underwings have a noticeable silver sheen and on the hindwings, a fairly broad orange band near the edge, next to a series of black spots, each centred with a bright, blue-green 'stud'.

Females are considerably duller. Upperwings are dark brown, sometimes with a hint of blue near the body, whilst underwing markings are similar to those of the male, but set against a deep brown background.

Remarkably, silver-studded blue butterflies enjoy a mutually beneficial relationship with ants – on heathland, these are most commonly the black ants *Lasius niger* and *Lasius alienus* – which in return for sweet secretions exuded by the caterpillars and pupae, offer a degree of protection to the butterfly in these, its early life stages.

Silver-studded blue butterflies mating in the sunshine

Southern damselflies – jewels in the New Forest crown

Tiny, needle-thin insects not quite 3 centimetres (1¼ inches) long and with a wingspan only marginally greater, mature male southern damselflies are

117

predominantly bright blue and black with a distinctive but variable black pattern, often referred to as a 'mercury' mark, on the second segment of the abdomen. Similarly sized females, some coloured predominantly green, others blue, have abdomens that are usually more darkly marked above than those of the males.

This southern damselfly had been wing-marked as part of a research project set up to examine the behaviour of these delicate insects

Out of water as winged adults from mid-May until early August, these nationally rather rare insects spend prolonged periods perched in the sunlight or, if the weather is particularly dull or windy, down amongst the herbage. Flight is weak, stuttering and usually at fairly low level.

Britain supports a significant proportion of the world's population even though here these largely sedentary insects have suffered a 30% decline in distribution during the last fifty years. Fragile creatures in more ways than one, southern damselflies have particularly demanding habitat requirements that inevitably make them vulnerable to change and also limit the number of places where they can successfully live. (On heathland, for example, it has been suggested that they prefer slightly acidic, shallow, reasonably well-vegetated, unpolluted streams with year-round water flowing slowly over a gravel or marl bed and with adjacent banks kept relatively free of encroaching scrub.)

But thankfully, southern damselflies continue to prosper in the New Forest.

More....
Dragonflies and damselflies – colourful, primarily wetland inhabitants: page 85
Beautiful demoiselles – incongruously exotic insects: page 137
Deadman Bottom: page 146
Dragonflies and damselflies – late fliers: page 185

Along the way

Golden-ringed dragonflies are often conspicuously present around the Crockford Stream and are also reasonably common in many places elsewhere within the New Forest. Enormous creatures with striking yellow bands that contrast sharply with their black background colouration, they fearlessly patrol the waterways whilst searching for smaller insects to devour.

Broad-bodied chaser dragonflies – quite large, wide bodied, pale blue males and similarly sized, primarily brown females – can also be seen around many of the ponds and streams, but they also wander widely onto heathland and will continue to do so until the end of their flight period in early August.

Look out, too, during this walk for a range of quite small damselflies, insects such as azure damselflies, blue-tailed damselflies, common blue damselflies, large red damselflies and southern damselflies; and for larger, beautiful demoiselles, downy emeralds, emperor dragonflies and keeled skimmers.

The Route

The first, preliminary detour

Golden-ringed dragonflies are often seen around the Crockford Stream

For those particularly interested in dragonflies and damselflies, this short detour from the Crockford car park to the Crockford Stream is likely, in season, to provide a range of excellent sightings.

Leave the Crockford car park at its western end – the end farthest from the road. Immediately go right, along a narrow heathland path; skirting on the right an area of woodland. After a short distance, cross a patch of open heathland where the woodland edge retreats to the right, and then pass beside an arm of protruding woodland. Go over another patch of open heathland and then follow the path as it bears a little to the left, downhill past further woodland, to reach the stream. Explorations to left and right here are unlikely to disappoint.

The walk can be rejoined by retracing the outward route back to the car park, or by following narrow, animal trackways diagonally across the heath to join the tracks used during Sections 1 and 2.

The main route

1. Leave the Crockford car park at its western end – the end farthest from the road. Follow a corridor of close-cropped grassland that almost immediately

119

> **Robins – silence reigns**
> In June, robins cease to sing, which really is a noteworthy event for these otherwise year-round choristers. Song – an attractive, fairly slow, always hesitant, intermittently sweet then plaintiff warble given by both males and females – will not resume until the second half of August or early September following completion of the adults' primary moult, and as young of the year finish the process of replacing speckled juvenile garb with adult plumage. All then are ready to use song and other often aggressive tactics to claim and defend autumn and winter territories.

bears a little to the right as a wide, grassy track; and pass minor paths to left and right.

2. After around 350 metres, turn right at a prominent 'T' junction where straight ahead, two very minor paths lead through the heather. Follow the track as, after another 150 metres, it bears close to 90 degrees to the left, ignoring on the corner minor paths to left and right.

3. After a further 300 metres, just after the track starts to bear left, turn right at a crossroads. Continue gradually downhill and after a short distance pass heathland paths to left and right. Follow the track as it bears a little to the right, and almost immediately reach the Crockford Stream – here on the left and right are extensive, usually very wet, areas popular with dragonflies and damselflies.
   Cross the stream using a makeshift, single-plank bridge put in place for this purpose.

4. **The second detour** takes walkers to a small pond, another potentially productive dragonfly and damselfly site. To take this detour, almost immediately turn left along a minor path running alongside what in late spring and summer is usually a very narrow stream, and after a short distance, where the course of the stream bears a little to the right, cross over the stream adjacent to a point where it widens to form what is usually

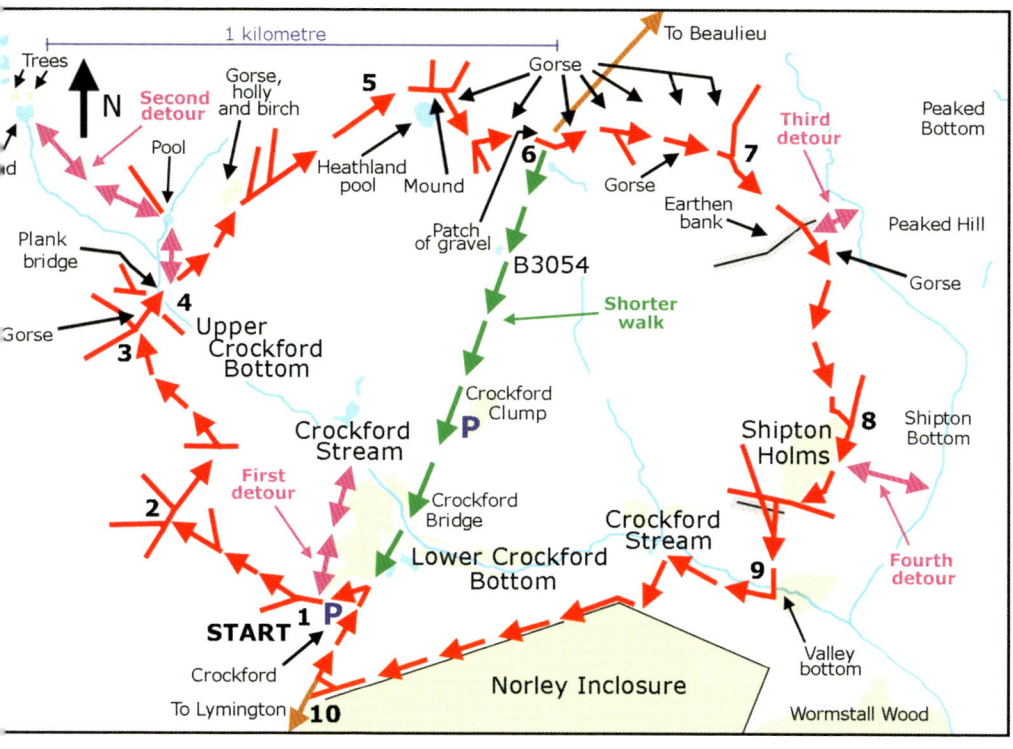

(in late spring and summer) a small, shallow pool. (There is no bridge here, but other than after heavy or prolonged rain, crossing is usually relatively straightforward.)

Immediately take the left fork at an indistinct 'Y' junction and walk along the side of a shallow, often wet valley on the left, heading towards two almost adjacent groups of trees visible in the mid-distance, beside which is the pond. When ready, retrace your steps to return to the main route.

Otherwise, after crossing the single-plank bridge, ignore the detour by continuing straight ahead along a quite wide, gravel track. After around 200 metres, skirt on the left a clump of gorse bushes and holly and birch trees; pass in quick succession two minor heathland paths on the left and go on straight ahead.

Notice to the right, on the horizon beyond the conifers of Norley Inclosure, the Isle of Wight's distant hills.

5. Follow the now dirt track as it eventually bears right, pass a path on the left and immediately skirt the edge of a large, man-made, gorse-clad mound with adjacent heathland pool that often dries out in the height of the summer.

The mound, associated with Beaulieu Heath's Second World War airfield, provided a backstop to targets used during the testing of aircraft machine guns.

Turn right, off the main track, immediately beyond and adjacent to the mound – the main track here bears left. Walk alongside, on the left, a patch of gorse and continue in the direction of the nearby minor road. After a short distance, ignore minor paths going straight ahead and half-right across the heath, and instead follow the now fairly inconspicuous path that bears left, running parallel to and not far from the edge of the gorse.

Pass on the left an irregularly shaped, quite small, bare area of ground strewn with gravel – possibly associated with the area's use as a Second World War airfield or maybe simply a shallow, dried out pool – and shortly after, reach on a bend the sometimes busy with traffic B3054, which runs from Beaulieu to Lymington.

Bog orchids – their pale green flowers are rarely noticed

Bog orchids grow in the dampest of places, often amongst floating carpets of sphagnum moss. Compared to many other orchids, they are inconspicuous plants that measure 3 -12 centimetres (1¼ - 4¾ inches) in height. Colonies in good years might exceptionally be two hundred strong.

Locally distributed in the south of the New Forest but scarcer in the north of the area, bog orchids nationally are rare plants. In fact, apart from those found in parts of the West Highlands of Scotland, the Forest has the largest concentration of populations in western Europe.

> ### Heathland wild flowers – two notable species
> **Lesser butterfly-orchids can be found in reasonably good numbers on and around a number of the heaths. Some populations grow on wet heaths; others below bracken on grassy heaths. Their delicate greenish-white blooms appear in June and July.**
>
>
>
>
> *Common dodder and bell heather* *Lesser butterfly-orchid*
>
> **Common dodder, meanwhile, a member of the bindweed family and a parasitic plant of gorse and heather, rarely grows abundantly, but is widely distributed. Tiny bell-shaped, pale pink flowers are produced and so are sprawling masses of slender reddened, thread-like stems that grasp for anchor points on the host plants, points that will also be used to extract nourishment from the hapless victims.**

6. **To take the shorter walk route**, turn right and follow beside the road back to the start point.

   Otherwise, cross the road and continue straight ahead for a very short distance beside on the left, the end of another quite substantial area of gorse. Almost immediately go half-left along a quite pronounced heathland track heading in the general direction of a distant tall chimney associated with Fawley Power Station – usually apparent, other than in conditions of poor visibility – with at the top a slender dark band above a paler, broader band.
   Almost immediately ignore a minor path on the right; continue straight ahead on a course broadly parallel to the edge of the gorse away to the

left; and then follow the track as it bears slightly to the right, past a small, path-side gorse clump on the right.

Continue down a modest incline and after a short distance go right at a junction where beyond, the land drops gradually away into a damp, shallow valley.

7. Go straight ahead in the general direction of a small wood a little to the right in the mid-distance – this is Shipton Holms, a broadly rectangular area of primarily holly and oak woodland. (Do not confuse this with the primarily coniferous Norley Inclosure still farther to the right.)

When around 200 metres from the previous junction, pass through a gap in a low, gorse and heather-clad earthen bank.

Late Bronze Age/early Iron Age enclosures in this area are overlaid by a more prominent complex of medieval enclosure boundary banks set down by the sheep farming monks of Beaulieu Abbey. (Shipton Holms was probably originally known as Sheeptown Holms.)

The third detour – immediately after the gap in the bank, a grassy path on the left leads down to a shallow stream that often dries out in summer, but is nevertheless worth checking out for dragonflies and damselflies.

Tit flocks form – a sure sign of summer

The breeding season over for another year, blue tits and great tits by mid-month are seen in loose, mobile flocks, sometimes accompanied on at least part of their travels by coal tits, chaffinches, goldcrests, long-tailed tits, nuthatches and treecreepers.

Great tit

(Flocking behaviour offers benefits but also has associated, presumably lower, costs. Flock members, for example, may be able to share sources of food discovered or disturbed by colleagues, and enjoy early warnings of predator presence provided by others within the flock. Competition for food is, however, likely to be increased (although birds of different species may feed on similar items but in non-competing ways) and socially subordinate flock members – often youngsters and females – may have their food stolen by more dominant individuals.)

Only with the approach of spring will the flocks no more be seen as members disperse and establish individual breeding territories anew.

Otherwise, continue straight ahead along a wider, grassy path running down a still modest incline alongside gorse on the left. Follow the path as it soon bears a little to the right before eventually going left, still downhill, heading towards the left-hand edge of Shipton Holms.

8. Reach the woodland edge, immediately pass a heathland path on the left, do not enter the wood, but instead continue clockwise around the edge of the wood.

   The fourth and final detour – here, immediately to the left, the ground drops rapidly downhill into Shipton Bottom: yet another potentially useful dragonfly/damselfly detour.

   Otherwise, when part-way round the wood, pass a minor path on the left and shortly after, as the wood on the right gives way to gorse and open heathland, take a fairly sharp left turn to follow a track running down into the wooded valley bottom – this track goes in the direction of the left-hand edge of Norley Inclosure on the hill-side opposite.

9. Cross the stream – there is no bridge, but in late spring and summer, other than after heavy or prolonged rain, the water is usually shallow – and immediately turn right, alongside the narrow ribbon of water. (Again, this is potentially productive dragonfly and damselfly habitat.)
   There are no prominent paths along here, but the route simply continues parallel to the stream. Eventually, after around 200 metres, as the stream-side land becomes increasingly wet and boggy, go left, uphill; and then turn right to follow the edge of Norley Inclosure back to the minor road, the B3054.

10. Turn right, alongside the road and the Crockford car park is a short distance away on the left.

For the adventurous, for those with a good sense of direction, strong map reading skills and access to an Ordnance Survey map!
Create your own walk by combining parts of this route with elements of your choice from the following selection of connecting or conveniently located nearby routes.

> From *New Forest Walks – a time traveller's guide:*
> Walk 10   Beaulieu Heath's Second World War airfield

Start	Small public car park beside the Happy Cheese pub, alongside the A35 in Ashurst – Ordnance Survey map reference SU335103. (Should space not be available here, alternative parking places can usually be found nearby)
Distance	5.5 kilometres (3½ miles) Shorter walk option: Reduce the distance by 1 km (0.6 miles) Walk extension: A detour onto heathland provides the potential to extend the route
Time to allow	1½ - 3½ hours
Refreshments	The New Forest and the Happy Cheese are both close to the start of the walk
Route	Along readily visible tracks
Terrain	Level ground with no significant gradients
Rating	1 – easy walking
Buggies	Suitable for sturdy buggies, but only if the shorter route is not used
Railway station	Ashurst (New Forest), adjacent to the start of the walk
Bus service	Bluestar
New Forest Tour Bus	Yes
Alternative start	None
'Camping in the Forest' Caravan and Campsite	Ashurst, 0.5 kilometres (⅓ mile)

JULY
Gentle rhythms dominate the natural world
(Ashurst: Churchplace Inclosure and Deerleap Inclosure)

The walk route passes through Deerleap Inclosure

The walk
Butterfly country! That would be a reasonably accurate description of some of the woodlands south-east of Ashurst through which this walk passes. Starting close to the Happy Cheese pub, the route follows a gravelled track leading to a keeper's cottage on the edge of Churchplace Inclosure, and goes

on through this inclosure and Deerleap Inclosure before returning along the same gravel track. A shortcut enables the 5.5 kilometre (3½ mile) route to be reduced by 1 kilometre (0.6 miles), whilst an optional extension is available onto adjacent heathland.

Featured wildlife

Hornets – fearsome reputations, yet usually placid insects

Although absent from much of Britain, hornets from May until October can frequently be seen flying along New Forest woodland rides. The largest European social wasp, they are considerably bigger than common wasps, but otherwise are of similar appearance. Stings are particularly venomous and if anybody is foolish enough to stir up the proverbial hornets' nest, they will no doubt regret it, but if left undisturbed, these impressive insects are usually placid, harmless creatures that ill-deserve their fearsome reputation.

Hornets are the largest European social wasp

Hornet flight, often laboured and clumsy, is accompanied by a loud hum caused by rapid beating of the wings. Diet is varied. Nectar is sometimes sought, but in spring and throughout much of summer, hornets prefer to take sugar-rich tree sap, and honeydew from leaves; whilst in late summer and autumn they also feed on apples, pears and other fruits. (Youngsters still in the nest – nests are made of wood pulp and are often located in a hollow tree – are fed primarily on pre-masticated day and night-flying insects.)

Only mated queens hibernate, and they alone survive the winter, ready in spring to found a new colony.

Roe deer – summer rut and spring births

The roe deer rut usually starts in late July and continues until mid-August. Roe deer rutting activity is, though, much less conspicuous than that of other New Forest deer and to casual observers can largely go unnoticed. However, during this time of high animal excitement, bucks can more regularly be seen by day as they purposefully follow after a doe or chase away other bucks, and may also be heard grunting or barking: in alarm, at a rival, to attract the attention of a doe or quite simply in testosterone-fuelled agitation.

Saplings stripped bare of bark by repeated antler action provide clues to the location of the rutting grounds and so do 'roe rings', well-trodden circular or figure-of-eight paths around trees, stumps or bushes. Often considered to be created during hectic rutting chases involving buck and doe, 'roe rings' do, however, sometimes show signs of winter use.

Roe deer have a true gestation period of around five months, but its onset after mating is delayed by about the same amount of time. Births consequently take place from mid-May until early June, at a time when copious cover is present to conceal the whereabouts of newborns, when food is usually readily available and when the weather is likely to be kind – young deer do not survive well in cold, wet conditions. Single births are most common on the Crown Lands, whilst in the wider countryside, in the absence of competition for food from larger deer and commoners' animals, twins are frequently born and triplets may also be encountered.

A roe buck in summer

More...
Roe deer – timid creatures of wood and heath: page 58

Silver-washed fritillaries – amongst the most handsome of butterflies

On bright July and August days, silver-washed fritillary butterflies are often on the wing in impressive numbers, seeking out bramble blossom and thistle flowers on which to feed. One of Britain's largest butterflies, they are common and widespread in some of the New Forest's broadleaved woodland inclosures where grazing pressure is limited and well-lit clearings, open rides or gravel tracks enable common dog-violets (the favoured larval foodplant), brambles and thistles to flourish – places such as Churchplace Inclosure, passed through during this walk.

Size alone is a useful aid to silver-washed fritillary identification and so, too, is their characteristically strong, swooping flight, an impressive sequence of graceful flaps and glides that carries them at speed over often considerable distances.

Look out, in particular, for the female variant known as valezina, a creature with pink on the under forewings and a dusky green background colour to the upper-wings, rather than the more usual deep orange colours of the male and the duller, brown-orange female shades. (These unusual insects so delighted the artist and naturalist F.W. Frohawk (1861-1946), who was a regular visitor to the New Forest, that he named his daughter Valezina in their honour.)

Silver-washed fritillaries are on the wing in July

Sweet chestnuts – summer catkins and autumnal nuts

Thought to have been introduced to Britain by the Romans, who valued their timber and edible nuts, sweet chestnut trees are reasonably widely distributed in the New Forest, but rarely do they grow in large numbers. Unmistakable, elongated, straw-coloured catkins in summer decorate the branches, producing a striking contrast with the lush, dark green foliage. Some – the longer catkins – are wholly male, whilst others – the shorter ones – have male flowers towards the tip and female flowers nearer the base.

Sweet chestnut leaves in all their autumnal glory

Although largely wind-pollinated, bees and other insects are regular visitors.

And whilst many catkins are destined to litter the ground beneath the boughs (for even moderate winds often seem to be enough to bring them tumbling down), the females that remain aloft will produce within prickly husks, chestnuts that by October will be rich-brown in colour, ripe and ready for eating.

Wild gladiolus – a New Forest speciality species

An illustrious member of the iris family, the wild gladiolus is one of the gems of the New Forest, and rightly so, for it occurs nowhere else in Britain. Distinctive, relatively tall hairless perennials with sword-shaped leaves, the plants are surprisingly difficult to see below their favoured canopy of bracken on grassy heaths. Prominent six-petalled, reddy purple flowers, pollinated mainly by large skipper butterflies, are present in June and July.

*Wild gladiolus:
a very special New Forest plant*

Wrens – tiny, cocked tail characters with secrets aplenty

Wrens are present in the New Forest throughout the year, often in relatively large numbers, although harsh winters take a severe toll on the health and survival rates of these tiny brown, mouse-like creatures. But despite their often

Wrens are more often heard than seen

confiding nature, a fondness for dense undergrowth reduces the likelihood of sightings. However, presence is frequently betrayed by unmistakably explosive, high volume song – a combination of loud notes and trills given in virtually all months of the year by both sexes, but most often by males – and by repeated, angry ticking calls and scolding, rapid churrs.

Many males are polygynous. Multiple domed nests, constructed of leaves, grass, bracken, moss and other items of vegetation, are often made by individual males in sites usually not too far from the ground – crevices in the root plates of upturned trees are often favoured. After choosing one for use, a female, usually unaided by the male, applies a lining of feathers and hair. Eggs are laid from late March or early April until June or July.

Communal roosting in harsh, winter weather (the birds cluster together to keep warm) has been recorded, but is infrequently observed. Otherwise, apart from during the breeding season, these often pugnacious birds live largely solitary, territorial lives.

Along the way

In addition to silver-washed fritillary butterflies in Churchplace Inclosure, look out for brimstones, commas, gatekeepers, holly blues, large skippers, large whites, meadow browns, painted ladies, peacocks, red admirals, ringlets, small heaths, small skippers, small whites and white admirals.

And for heathland variety, be sure to investigate the route extension available mid-way round – at the beginning of Section 6.

White admirals can be seen in Churchplace Inclosure

The Route

If joining the route from the railway station, leave the station at the end of the platform serving trains going towards Brockenhurst and Christchurch, and walk along a fence-side tarmac path signed 'Way out to Ashurst Village'. Go through a gate and follow the path as it bends to the right, just before a railway bridge. Pass another gate, on the right; and reach the public car park – the walk start point.

1. Leave the car park on the side close to the Happy Cheese pub, pass a road on the right leading into the NHS Ashurst Centre – once the site of the New Forest Union workhouse – and follow the tarmac road running half-right, behind the Happy Cheese.

After a short distance, go through a gate; pass a number of clumps of butcher's broom – a prickly, evergreen shrub – on the left; and continue straight ahead along a gravel track with fields on the left, and on the right,

> **Fungi of the summer woods – two species: one edible, the other not**
>
>
>
>
>
> *Chicken of the woods*
>
> **Chicken of the woods is a conspicuous fungus that, as the name suggests, is extremely edible. A reasonably common species, its often quite large, succulent fruiting bodies – tiered clusters of fibrous, sulphur-yellow, bracket-like plates – are said to remind of chicken meat. If eaten when young, they provide a tasty addition to soups and stews, whilst slices can be grilled or used as cutlets.**
>
> *Ganoderma adspersum*
>
> ***Ganoderma adspersum*, meanwhile, is a large, perennial, inedible, common and widespread bracket-like fungus that most frequently grows on beech and oak trees. It, too, is conspicuous and in summer sheds rust-red spores that add prominent patches of colour to the host trunk and ground below.**
>
> **More....**
> **Woodland and other fungi – a harvest to be taken in moderation and with care: page 178**
> **Beech trees – obliging fungal hosts: page 191**

133

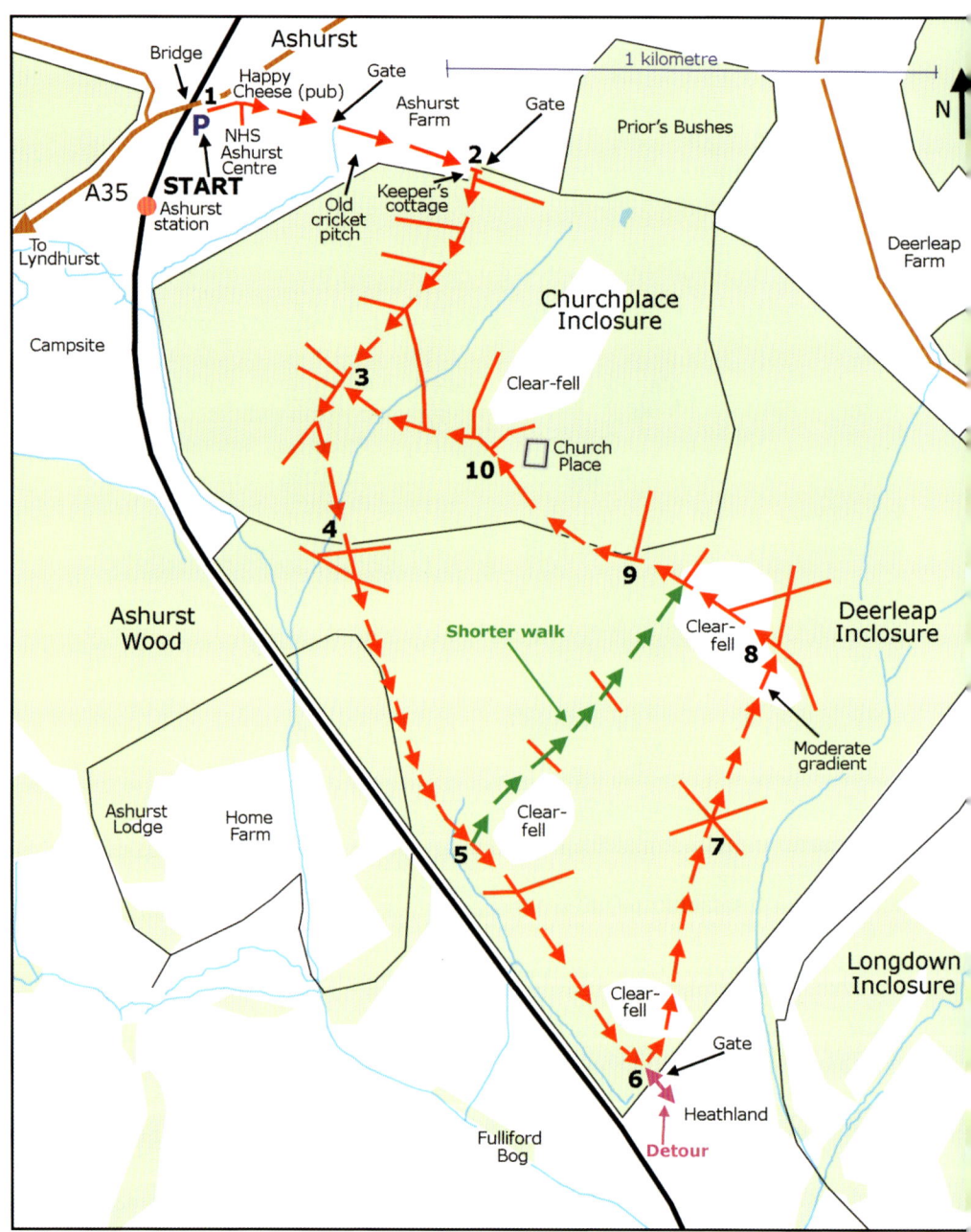

a small New Forest lawn that until relatively recently was used as a cricket pitch – on match days a sign was displayed asking walkers to wait for a break in the play before passing!

Continue beside, on the right, an attractive, presumably very old, hedgerow containing holly, ash, hawthorn, oak, hazel, and yew; with dog rose and honeysuckle haphazardly sprawling over the top.

2. Pass a keeper's cottage – Church Place Cottage – on the right; turn right, through an adjacent gate leading into Churchplace Inclosure; ignore a path immediately on the left and continue straight ahead along a gravel track leading into this attractive, predominantly broadleaved woodland.

Ignore over the next 450 metres, or so, three woodland rides on the right of the main track and another on the left.

Numerous mature oaks have been felled along here, benefitting those that remain by creating space and light for further growth. Bramble growth, too, will be encouraged, much to the benefit of butterflies and other insects.

3. At the next staggered junction of tracks, continue straight ahead along what here becomes a cycle

Wood sage – a plant that usually goes unnoticed

A rather dowdy, shrubby plant that is often overlooked, wood sage occurs around heathland edges and in New Forest woodlands, including Churchplace Inclosure. Its tough, sturdy, predominantly erect stems can grow to a height of 50 centimetres (almost 20 inches), whilst the flowers, present from June to September, are usually a greenish-yellow colour, but sometimes white, and conspicuously lipped.

Although now neglected by man, in former times, wood sage was used to flavour and preserve beer and was also thought to be effective as a diuretic, for healing cuts and other open wounds, and to combat ulcers, scurvy and rheumatism. Country names such as gipsy's baccy and gipsy's sage hint also at a traditional association with Romany folk.

The delicate blooms of wood sage are best appreciated in close-up when the prominent, maroon stamens can be clearly seen

track; and shortly after, pass a narrow track on the right as the main track bears round to the left.

4. Pass through a gap in the wood-bank separating Churchplace Inclosure from Deerleap Inclosure – the wood-bank is barely noticeable when engulfed in summer vegetation – and immediately after, go straight ahead at a relatively inconspicuous, minor crossroads.

After a further short distance, continue straight ahead at a second crossroads – this crossroads is more prominent than the last, although there is only an overgrown path to the left, whilst another cycle track goes to the right.

Follow the track as it bears a little to the left and when 600 metres beyond the previous crossroads, immediately after the track again bears left, reach a quite wide, grassy ride on the left.

5. **To take the shorter walk route** (which misses out the south-eastern section of the route), turn left here and then continue straight ahead – nearby on the right after a short distance is a quite extensive area of regenerating clear-fell.

(This secluded, little-used ride quickly opens out to let in lots of

> **Stag beetles – pugnacious males equipped for battle**
> Although nationally rather scarce, stag beetles are reasonably common and widespread in the New Forest. The longest of the UK beetle tribe, males measure up to 7.5 centimetres (3 inches) in length, excluding their rather exotic, stag-like mandibles that are used to good effect as warning signals to competing males and as weapons during battle. Female body length can reach a still-impressive 4.5 centimetres (almost 2 inches).
>
> Much of their life is spent as rather unpleasant looking, primarily cream coloured larvae that feed on rotting wood before pupation within a cocoon, specially constructed for the purpose, hidden away quite deep in the soil.
>
> From late May to the end of August, however, winged adults are seen. Many are attracted to house lights and street lamps, and can sometimes be found on pavements below the lamps – a characteristic shared with May bugs, the equally robust and impressive cockchafer beetles that are also a prominent feature of New Forest life.

Male stag beetles are equipped with enormous 'antlers'

light, and is just the sort of out-of-the-way place where the quiet, observant walker can expect to encounter roe, fallow and occasional muntjac deer, particularly early and late in the day.)

Go straight ahead at two minor crossroads and, 700 metres after the shorter walk route was taken, rejoin the main route towards the end of Section 8 by turning left at a crossroads, along a quite wide, gravel track.

Otherwise, continue straight ahead and, after a relatively short distance, go straight ahead again at another crossroads – there is a quite wide, gravel track to the left and a grassy track to the right.

6. Eventually follow the cycle track as it goes sharply round to the left.

To extend the route and add variety to this otherwise predominantly woodland walk, follow a track on the right at this corner; go through a gate and enter an attractive, extensive area of heathland. Wander at leisure before retracing your steps back through the gate.

Otherwise, continue along the cycle track, initially across a narrow strip of recently clear-felled land.

Beautiful demoiselles – incongruously exotic insects

Beautiful demoiselle damselflies are typically found around quite fast flowing, often gravel-bottomed streams, although they sometimes wander into woodland. First seen on the wing towards the end of May, their flight period lasts until late August. Males, frequently strongly territorial creatures, are a bold, metallic blue-green colour, whilst females have golden brown wings with metallic green present on the head, thorax and abdomen.

More....
Dragonflies and damselflies – colourful, primarily wetland inhabitants: page 85
Southern damselflies – jewels in the New Forest crown: page 117
The Crockford Stream: page 118
Deadman Bottom: page 146
Dragonflies and damselflies – late fliers: page 185

7. Eventually continue straight ahead at the next junction of tracks; meet an extensive area of regenerating clear-fell; and, after walking up a hill of fairly moderate gradient, take the next left turn, at a crossroads, along another gravel cycle track.

8. Go down a short hill; pass a narrow path on the right, largely obscured by vegetation; and eventually go straight ahead at a crossroads where the cycle track intersects the grassy ride used during the shorter walk route.

9. Pass a track on the right and continue along the cycle track as it first bears left, downhill, and then right, up a gentle gradient. Leave the conifers of Deerleap Inclosure and re-enter the broadleaved woodland of Churchplace Inclosure – on the left, running at a sharp angle to the cycle track, the moss-encrusted earthen bank separating the two inclosures is visible.

10. As the track again goes gently downhill, on the right is a bracken-clad knoll topped with mature oak, beech and holly trees.
    Known as Church Place, this is the site of a medieval keeper's lodge, the name of which is of similar origin to The Churchyard at Sloden, passed during the December 'Fritham: Eyeworth Pond' walk.
    Shortly after, as the main track bears left, pass a track on the right and continue gradually downhill. Immediately pass on the right, a grassy track that becomes quite overgrown in summer; and soon after, another, also on the right. Eventually, at the staggered junction of tracks passed at the beginning of Section 3, turn right and retrace the route back to the car park (and station).

For the adventurous, for those with a good sense of direction, strong map reading skills and access to an Ordnance Survey map!
Create your own walk by combining parts of this route with elements of your choice from the following selection of connecting or conveniently located nearby routes.

From *New Forest Walks – a time traveller's guide:*
Walk 6   Ashurst: Busketts Inclosure
Walk 7   Ashurst: Churchplace Inclosure

AUGUST
Heathlands pink with heather blossom
(Ashley Walk, Millersford Bottom, Godshill Inclosure, Hale Purlieu, Millersford Plantation, Deadman Bottom and Cunninger Bottom)

Summer heathland: a magnificent sight

The walk
This heathland, woodland and wetland walk requires a moderate level of fitness for comfortable completion, but the rewards make the effort worthwhile: panoramic views over meandering valleys, cooling shade provided by attractive woodlands, and the presence of narrow streams and areas of

Start	Ashley Walk, Forestry Commission car park 1.25 kilometres (¾ mile) north-east of Godshill on the B3078 Brook to Fordingbridge road – Ordnance Survey map reference SU186156
Distance	11.5 kilometres (7¼ miles) Shorter walk options: 1) Reduce the distance by 0.25 km (⅙ mile) 2) Reduce the distance by 6.5 km (4 miles) 3) Reduce the distance by 6 km (3¾ miles) 4) Reduce the distance by 0.5 km (⅓ mile)
Time to allow	3 - 7¼ hours
Refreshments	The Fighting Cocks, Godshill, is close to the start of the walk
Route	Largely along readily visible tracks, although in places – for example, during Sections 7 and 8 – a little 'off the beaten track'
Terrain	Undulating ground with a number of moderate gradients. Note: After heavy rain and in winter, the ground in places – for example, during parts of Sections 16, 17 and 18 – can be rather wet. The use of waterproof footwear is particularly recommended at these times
Rating	3 – in places, quite strenuous walking
Buggies	Not suitable
Railway station	Ashurst (New Forest), 19 kilometres (12 miles)
Bus service	More/Wilts and Dorset serve nearby Fordingbridge
New Forest Tour Bus	Yes
Alternative starts	1) Deadman Hill, Forestry Commission car park at Ordnance Survey map reference SU192165 2) Godshill Cricket, Forestry Commission car park at Ordnance Survey map reference SU182151 3) Turf Hill, Forestry Commission car park at Ordnance Survey map reference SU212177 4) Godshill, Forestry Commission car park at Ordnance Survey map reference SU177161
'Camping in the Forest' Caravan and Campsites	1) Longbeech, 10 kilometres (6¼ miles) 2) Ocknell, 11 kilometres (7 miles)

wetland much loved by dragonflies, damselflies and a variety of breeding birds. Those with more modest energy levels or little inclination for exertion need not, however, despair, for short-cuts are available that reduce the overall distance of 11.5 kilometres (7¼ miles) by up to 7.25 kilometres (4½ miles).

Featured wildlife

Bog asphodel – bright yellow mid-summer blooms

Bog asphodel blooms appear from mid-June and are at their best for much of July. Common and widespread in damper parts of the New Forest, bog asphodels are conspicuous, erect plants that grow to a height of up to 40 centimetres (16 inches), often in loose colonies amongst grasses and other wetland vegetation. Six-petalled, bright yellow flowers grow in spikes towards the tip of the stems. By August, many blooms will have taken on a deep orange colour, whilst others will have already given way to reddish-orange, egg-shaped fruits containing tiny seeds, each seed with a tail-like appendage at either end designed by nature to increase buoyancy – so important should they fall into water.

Bog asphodel blooms illuminate the wetter places

When eaten, the plant in days-gone-by had a reputation for weakening the bones of sheep and cattle – part of the Latin name refers to bone-breaking. This old belief has, however, been disproved, no doubt much to the relief of New Forest commoners whose stock consume just about anything edible that takes their fancy.

Crab apples – late summer and autumnal feasts

A tree of modest height, often as broad as tall, and with trunk and limbs crooked and twisted, the crab apple is certainly distinctive. Frequently blessed with thorns aplenty and fruits that are small, bitter and hard, these humble natives

141

are the proud ancestors of modern, cultivated apple varieties created by horticulturists during centuries of selective breeding and improvement.

In May, blizzards of crab apple blossom, strikingly rose-pink and white, light up New Forest heathlands, grasslands and open woodlands. Fruits are well-formed by August and by October have mostly fallen, although the more stubborn

Crab apples hang aloft, not yet ready to fall to earth

might remain aloft, clinging to soon-to-be-bare branches. Birds gorge on the ripe windfalls and so do a range of mammals, whilst hornets, commas and red admiral butterflies feast on the juices. Commoners' ponies, however, somewhat perversely often seem to prefer to feed at this time on potentially poisonous acorns.

Green woodpeckers – ever-noisy yaffles

The loud, far-carrying, laughing cries of green woodpeckers can be heard in most months of the year, but they are used with increased frequency by both sexes during the early stages of the breeding cycle – from late March until early May. But unlike great and lesser spotted woodpeckers, green woodpeckers rarely drum.

Also known as the yaffle, this green and yellow torpedo-shaped bird with

Green woodpeckers are sometimes known as yaffles

conspicuously bounding flight was considered by country people to be a herald of rain, probably because its call is better heard in the clear atmosphere that often precedes a downpour – another old, country name was rain-bird.

Relatively common and widespread in the New Forest, green woodpeckers are most often seen around woodland edges and on lightly wooded heaths, rather than in dense woodland. Sound most often betrays presence, whilst a typical sighting is of a rapidly departing bird flying away after being disturbed, landing in upright posture on a distant birch trunk, sidling round, out of sight of onlookers and then peeping back to investigate the cause of alarm.

Heathers – rippling waves of pink and purple

Few can fail to be impressed by the glorious sight of wide expanses of New Forest heathland glowing pink and purple, brought to life by the flowering of the heathers.

Of the three common species present, heather, or ling, as it is sometimes known, occurs most abundantly. A long-lived, wiry, sprawling plant, heather's much-branched tangled stems retain a year-round wine-red colour, whilst the loose spikes of delicate, usually pinkish-purple flowers are best seen from mid-August until mid-September.

The other two species – bell heather and cross-leaved heath – are superficially similar to heather, but are far less likely to monopolise extensive tracts of land. In fact, both are quite selective about where they grow. Bell heather favours relatively dry ground where it shows off dark green foliage

A honey bee foraging on heather

Bell heather shows off its bell-shaped flowers

and robust, usually reddish-purple bell-shaped flowers; whilst cross-leaved heath prefers to grace damper places with its downy, silvery grey-green foliage and globular, usually pale pink flowers. Both bloom from June until September or early October.

Rich in nectar and pollen, all attract a wide variety of insects to feed, including honeybees that can often be seen buzzing from plant to plant, thighs and other body parts heavily loaded with the summer's harvest.

During autumn and winter, as grass becomes progressively less readily available, young heather shoots form an increasingly important element of the diet of ponies and other stock animals. But heather as it ages, becomes less palatable, so relatively small sections of heathland are burnt on a controlled, rotational basis to eliminate tough old stems and encourage the regeneration of tasty young replacements (along with fresh new grasses and gorse shoots) whilst at the same time removing, or at least setting back, the growth of invasive trees and other undesirable vegetation that might otherwise encourage natural succession to woodland.

Pale pink flowers of cross-leaved heath

A kestrel hovers in the breeze

Kestrels – aptly named hoverhawks

Kestrels, medium sized, rather distinctive falcons, are present in the New Forest right the way through the year and can sometimes be seen hovering aloft, staring intently downwards, watching for movements that signal the presence of a small mammal or other prey item. Indeed, in mid- and late summer, kestrel families in well-spread groups may be encountered, the youngsters of the year and their parents all hanging in the sky within sight of each other as the inexperienced birds learn to hunt for themselves. (Their frequently observed hovering habits, not surprisingly, gave rise to country names such as hoverhawk, wind cutter, windhover, wind fanner and windsucker.) Numbers on the Crown Lands are, however, relatively low, particularly in winter when some move away in search of easier hunting elsewhere.

Young are often raised in old, carrion crows' nests situated high in Scots pines or other coniferous trees. Between three to six eggs are usually laid in early April, or sometimes later. Incubation lasts for around one month, the young fledge after a further month and remain at least partially dependent upon the adults for a minimum of another month. (Small grey pellets, the regurgitated indigestible remains of prey items – fur, bones, beetle wingcases, for example – may sometimes be found below nest sites and regularly used perches.)

Reptiles – late summer youngsters

Following April and May matings, female adders in August and September produce live, fully formed young. Litters range in size from three to twenty, but average around ten. Pencil-thin creatures up to 20 centimetres (8 inches) long, young adders may remain with the female for a few days before dispersing to make their own way in the world. Initially greyish in colour, they soon take on dark brown tones. Maturity is reached after around three years. Similarly, following spring-time emergence from hibernation and subsequent mating, common lizards, slow worms and smooth snakes also give birth to live young in late summer or early autumn.

Two young adders rest with 'mum', a strikingly black creature

Grass snakes, meanwhile, produce eggs, usually in June or July, that are frequently placed in rotting vegetation – compost heaps are favourite sites – capable of acting as an incubator, whilst sand lizards lay eggs in late May or early June and usually bury them in sand exposed to the sun. Largely left to their fate, unless predated or otherwise damaged, the eggs of both species hatch in late summer or early autumn.

More...
Reptiles – far better behaved than their reputation suggests: page 103

145

Sundews – ferocious wetland plants

Surely amongst the most curious of New Forest plants, sundews supplement the meagre nutrients available in their favoured wetland soils by absorbing the body parts of insect prey. Three closely related species occur: round-leaved sundews and oblong-leaved sundews are locally common, but great sundews are far scarcer. (The first two can be differentiated by leaf shape, the third by size.)

Round-leaved sundews trap unwary insects

Found on many of the wetter heaths and around the margins of bogs and mires, these perennial plants are all of relatively modest size. Un-branched stems from June to August bear fairly inconspicuous, tiny white flowers, whilst the leaves boast deadly arsenals of reddish coloured hairs, each tipped with a sticky, swollen gland that, just like dew, glistens in the sun. Upon contact with the sticky tip, insects as large as damselflies and craneflies quickly become ensnared; other hairs around the point of contact gradually bend inwards, effectively preventing escape; and digestive secretions produced by the plant dissolve the softer parts of the insect's body before the resultant liquid is absorbed by the ever-voracious predator.

Along the way

Managed by the National Trust, Hale Purlieu features a notable mixture of heathland and boggy mires that attract a wide range of wildlife, including Dartford warblers, stonechats, nightjars, hobbies, curlews, snipe, fallow deer and roe deer. The pond in Deadman Bottom and the adjacent area of marshy land also often repay exploration in search of dragonflies and damselflies. Insects present might include beautiful demoiselles, common darters, emperor dragonflies, four-spotted chasers, golden-ringed dragonflies, keeled skimmers and southern hawkers.

A summer's day on Hale Purlieu

The Route

1. Leave the Ashley Walk car park by its main entrance, cross the adjacent road, turn right and continue along the strip of grass between the road and nearby fence. After a short distance, follow the fence-line as it goes close to 90 degrees to the left; and eventually continue downhill, along a path running parallel to the fence, through gorse and other vegetation.

2. Pass a gate on the left with a public footpath beyond; continue downhill, parallel to the fence and bank; reach a valley-bottom stream fringed by trees; turn right and then immediately left to cross the stream using a railed bridge.

3. Immediately go left and then right to regain the route's original course, and continue steadily uphill along a quite wide track.

   Soon leave behind the left-hand fence and bank; immediately ignore a path on the left; continue straight ahead, still uphill; and eventually reach a 'Y' junction on a quite steep section of the hill.

   To use the first shorter walk route (which misses out Godshill Inclosure and provides extensive views over Millersford Bottom), take the right-hand fork; after a relatively short distance, turn right at the next 'T' junction; and follow what initially is a very pronounced gravel track running broadly parallel to, and 100 metres, or so, from, the boundary of Godshill Inclosure.

   Continue straight ahead on this meandering hill-top course for almost 1 kilometre (0.6 miles) before eventually rejoining the main route adjacent to the quite large property encountered at the end of Section 5.

Rowan trees – spring flowers and late summer berries

Rowan trees in the New Forest are most often noticed on the heaths, particularly in late spring when clusters of creamy-white flowers engulf the branches, and again in August when scarlet berries provide feasts for blackbirds, song thrushes and mistle thrushes. Indeed, so irresistible are the berries, that bird-catchers of old frequently used them as bait when trapping unwary victims.

147

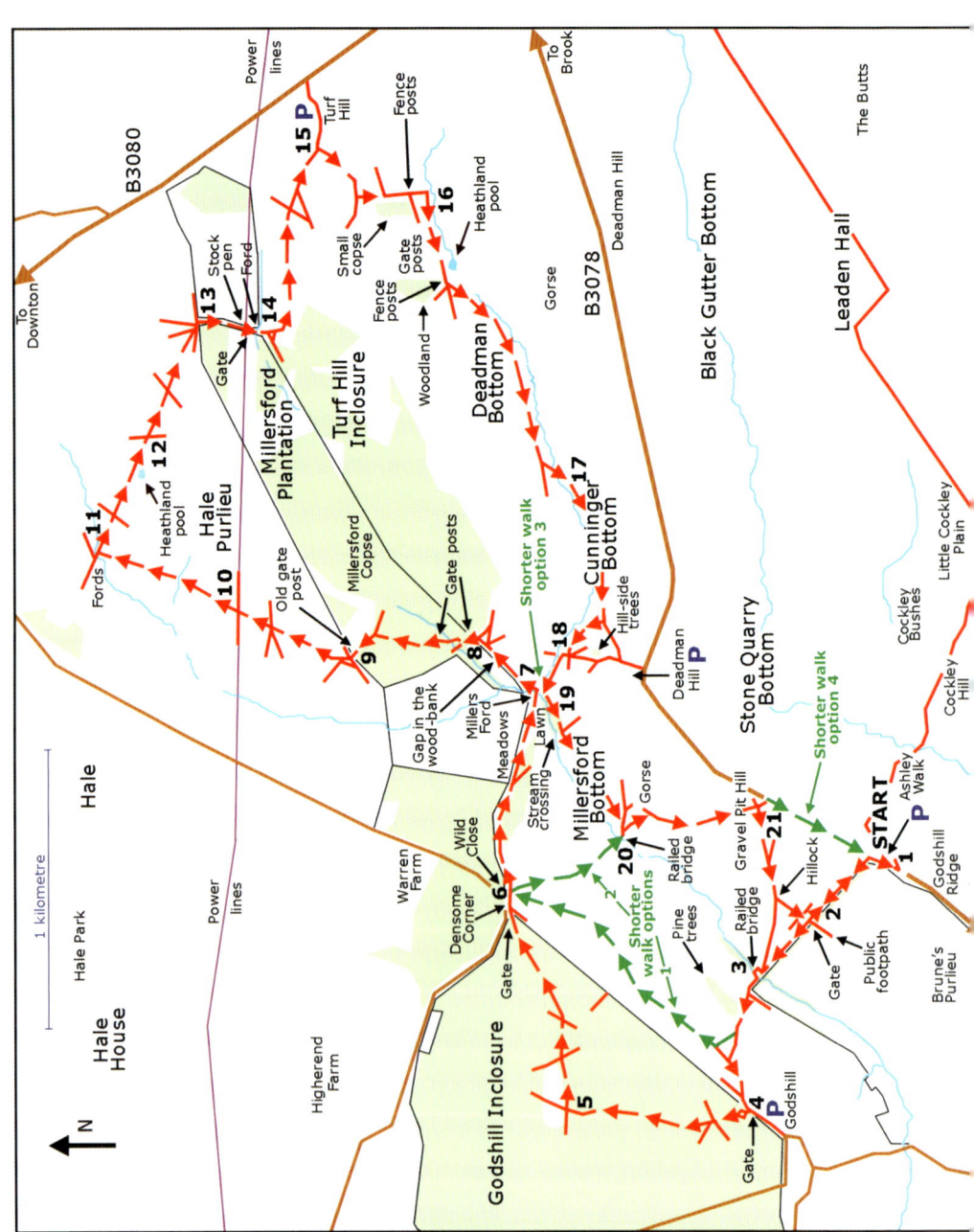

Otherwise, take the left-hand fork at the 'Y' junction and then as the gradient decreases, follow the track as it bears a little to the right and after a short distance, crosses a dip in the land.

Go left at the next 'T' junction to reach after a short distance, the Godshill, Forestry Commission car park.

4. Pass beside a low, vehicle barrier on the edge of the car park; immediately enter Godshill Inclosure through a pedestrian gate on the right; and straightaway take either the left-hand or right-hand fork in the path – they rejoin a little farther on.

Go straight across a gravel track, immediately take the right-hand fork at a 'Y' junction and continue straight ahead along a quite pronounced track.

5. At the next crossroads, turn right, along a substantial gravel track. Continue straight ahead at a less pronounced crossroads and again where a gravel track intersects the route – follow here a narrower path that first bears a little left and eventually leaves the inclosure through a pedestrian gate close to Densome Corner.

Turn right and continue along the near side of an adjacent minor road. Almost immediately – as the road bears quite sharply left – continue straight ahead past a Forestry Commission vehicle barrier, and along a gravel track beside which, set back on the left, is a quite large property called Wild Close.

6. **To take the second shorter walk route** (which cuts out the entire north-eastern section of the walk), turn half-right directly opposite this property, go across a small, grassy clearing and follow the path diagonally downhill. Cross a railed bridge over a valley-bottom stream; go straight ahead for a very short distance, ignoring here tracks to the left; and continue along the main route from Section 20.

Otherwise, continue downhill, parallel to the left-hand woodland edge, at times within a narrow strip of woodland, at other times through woodland clearings. (Not far away to the right, wet, boggy sloping ground leads down into Millersford Bottom; whilst on the left, a number of gates give access to the private lands beyond.)

Eventually leave behind the left-hand woodland and continue straight ahead, close to, and parallel with, the left-hand fence which now borders privately owned meadows.

Emerge onto a quite large, close-cropped lawn bordered by, and interspersed with, hawthorns, blackthorns, brambles and gorse; and follow the grassland edge on a course still close to, and broadly parallel with, the left-hand fence-line.

At Millers Ford, cross a valley-bottom stream which for much of the year can readily be stepped over, although heavy or prolonged rain might, as ever, cause some flooding.

7. **To take the third shorter walk route** (which misses out much of the north-eastern section of the walk), immediately turn right and follow the main route directions from Section 19.

Otherwise, bear left alongside the woodland/heathland edge, ignore minor paths to left and right, pass a quite wide gap in the wood-bank on the left and follow the path straight ahead below over-shading Scots pines.

8. Shortly after, reach a 'T' junction where straight ahead is dense woodland; and turn left, downhill, along a quite wide gravel/dirt track.

Immediately go between two old gate posts – and through a gap in the wood-bank – to enter plantation woodland. Pass beside on the left, a narrow corridor of clear-felled land invaded now by encroaching birch; immediately take the right-hand, more substantial fork at a 'Y' junction; and follow the track as after a short distance it bears left.

Cross a usually narrow stream at a hollow worn by passing hooves and feet – there is no bridge here. Continue half-to-the right between further old gate posts and go on uphill.

After a short distance, follow the track as it again bears slightly right; then as the ground begins temporarily to level out, ignore a track coming in on the right to form an inverted 'Y' junction; and continue half-left in the direction of power lines visible in the mid-distance.

9. Follow the track through a gap in the woodland boundary bank where on the right is a solid, old wooden gate post; immediately ignore on the left, a path running downhill, parallel to the woodland edge; and ignore on the right, a very minor path running uphill, again broadly parallel to the woodland edge.

Continue along a now fairly prominent gravel track running out across the National Trust heathland of Hale Purlieu; and almost immediately follow this as it bears right, uphill, ignoring here further tracks going off to the left and straight ahead.

Grayling butterflies – widespread residents

Primarily on the wing from mid-July until mid-September, grayling butterflies in the New Forest are most often seen on dry heathlands where they pilfer nectar from the heathers. Boldly marked, light brown and straw coloured upperwings are glimpsed only as they speed by in often erratic flight, for when at rest the wings are almost always held tightly shut. (In warm weather, wing edges are invariably tilted towards the sun, minimising the amount of sunlight hitting the wings and maybe helping the butterflies to avoid over-heating. Or maybe this behaviour is a camouflage tactic designed to reduce the amount of shadow cast.)

After bearing right, continue straight ahead towards the power lines. At the top of the hill, go straight ahead at a junction and go straight ahead again at a crossroads beneath the power lines.

10. Proceed gradually downhill; cross a narrow, shallow valley with an often-dry stream piped under the track; eventually go down a further short hill towards a valley-bottom stream – also often dry in summer – and turn right immediately before the stream.

11. Continue gradually uphill along a quite pronounced, gravelly heathland track; after a short distance, go straight ahead at a quite indistinct crossroads; and as the ground levels out, pass on the right, a small heathland pool.

12. Eventually go straight ahead at a crossroads of wide, grassy tracks and again at another crossroads. Cross a narrow, shallow valley; eventually ignore tracks to the left and right; and instead, follow the main track as it bears right, downhill, between two fenced blocks of land.

13. Continue downhill and pass through the gates of a quite large, stock holding pen located beneath the power lines. Immediately after, pass a gate on the right; continue downhill into a narrow, wooded valley and cross a small stream – often dry in summer – at a ford.

151

14. Immediately take the left-hand fork at a junction, go uphill for a short distance and as the ground largely levels out, follow the main, fairly wide gravel track as it bears left across open heathland.

Continue gradually uphill and follow the track as it eventually bears a little to the right, goes over an indistinct crossroads and passes another, more prominent track on the right.

15. After a short distance (immediately before the track starts to bear a little to the left, not far from the trees around the Turf Hill car park), follow along on the right, a wide corridor of close-cropped grassland.

The land around this car park was previously the site of administration, storage and preparation facilities associated with the Second World War, Armaments Research Department, Millersford. The operational area, where explosives were tested and the results recorded, was on the ridge of Turf Hill around 800 metres away to the west-south-west.

Continue along the left-hand edge of the grassland corridor. After around 200 metres, as the corridor starts to bear round to the right, follow an equally wide grassy corridor to the left. (This eventually runs downhill to Deadman Bottom. On the right

Ravens – highly intelligent members of the crow family

Ravens, particularly in summer, often in pairs or small groups, can sometimes be seen and heard squabbling, displaying high in the sky, engaged in bouts of conversational vocalisation or simply perched on the pylons that cross the area passed through during this walk. Relatively recent breeding re-colonists of the New Forest after an absence of almost 150 years, these enormous black members of the crow family often first draw attention to themselves with loud, unmistakable, far-carrying croaks.

Thought to pair for life, ravens exhibit many sophisticated behavioural traits including, it is said, the ability for mates to recognise one another over considerable distances and communicate modified vocal information intended only for each other.

A newly fledged raven cries out to be fed

152

along here, after a short distance, is a small copse.)

When part-way down the hill, ignore a wide, grassy track on the left; and a little farther on, pass a path on the right bordering the southern edge of the copse. Also pass by fence posts on the left.

Immediately before a stream is reached, turn right, along a grassy path running close to the valley bottom.

16. Follow the path on its course close to the stream; after a relatively short distance, pass a pair of old, isolated, hill-side gate posts on the right; and immediately after, reach a small heathland pool on the left, with beyond, a very wet area of boggy ground – ideal dragonfly and damselfly habitat, a watery oasis in an otherwise often dry, summer landscape.

Almost immediately pass on the right, two more old wooden posts and a narrow strip of woodland. Immediately go half-left at a junction of tracks and continue on a course running broadly parallel to, and not far from, the valley bottom.

Pass quite extensive areas of hill-side gorse away to the left; and when around 800 metres from the previous junction, a little before the track bears right, walk over a narrow area of grassland on the left and cross the valley bottom – the stream is often completely dry here in summer, its course difficult to discern.

17. Turn right; continue along a quite wide, grassy path running along the bottom of the hill-side on the left; and eventually – as the main path starts to go quite steeply uphill to the left, towards the Deadman Hill car park – go right, and continue along a minor path running close to the base of the hill.

18. After around 200 metres – a little beyond a group of hill-side trees on the left – where the route is joined firstly by a narrow track on the left and then by a second, more prominent track running downhill from the direction of the Deadman Hill car park, turn right, along the second track, and go downhill towards the valley bottom.

After a short distance, go left immediately before what in summer is usually a very narrow stream – the same stream that runs intermittently along Cunninger Bottom and Deadman Bottom. After a further short distance, go left again, immediately before Millers Ford and the more substantial stream crossed on the outward route - at the end of Section 6.

19. Continue along a narrow path running through trees and bracken adjacent

> **Dwarf gorse – heathland ankle nipper**
> Dwarf gorse adds splashes of bold yellow, ground level colour to the heaths, providing a vivid contrast with the pink and purple heathers. In bloom from mid-summer until early autumn, it is endowed with plentiful spines that can inflict sharp pain to uncovered lower legs and ankles. (As the name suggests, this is a low growing plant – although the stems can be quite long, they often grow out from the ground at a shallow angle.)
>
>
>
> More....
> Gorse – colourfully widespread furze: page 64

to the stream. After a further short distance, pass another stream crossing place - used by ponies, cattle and other animals - on the edge of open heathland, with here a minor path on the left.

Go almost straight ahead along what initially is a quite wide, grassy track running across the heath, a little away from the stream. Ignore along the way, a number of minor paths to left and right.

Eventually reach a railed bridge carrying the shorter route from Densome Corner (detailed at the start of Section 6), and when adjacent to the bridge, go 90 degrees to the left.

20. Follow a very minor path, running through a narrow patch of gorse, in the direction of a nearby, more substantial gravel track leading diagonally uphill, across the heath to the right.

After around 100 metres, when close to the edge of an area of hill-side gorse, turn half-right to follow the more substantial gravel track uphill on a course leading towards the nearby ridge-top road – the B3078.

21. **To take the fourth and final shorter walk route** (which cuts out the remaining heathland section of the walk), turn right, alongside this road to return to the start of the walk.

Otherwise, when near to the top of the hill – around 60 metres before the road is reached – turn right at a crossroads of very minor paths, and continue along a narrow, meandering heathland path that initially runs just below the hill-top. Follow this as it bears slightly right, pass a minor path coming in from the left, and continue predominantly down a gentle hill along what here is a fairly wide, undulating gravel track.

Ignore minor paths to left and right; and at the top of the last, rather modest-sized, bracken, gorse and heather-clad hillock (from where the main track continues downhill towards the stream), go half-to-the-left along a minor path.

Continue straight ahead at a crossroads; immediately after, turn left close to the fence, wood-bank, line of trees, gate and public footpath passed on the outward route; go uphill back to the minor road and turn right to reach the Ashley Walk car park.

For the adventurous, for those with a good sense of direction, strong map reading skills and access to an Ordnance Survey map!
Create your own walk by combining parts of this route with elements of your choice from the following selection of connecting or conveniently located nearby routes.

From *New Forest Walks – a seasonal wildlife guide:*
January   Godshill Cricket

From *New Forest Walks – a time traveller's guide:*
Walk 1   Ashley Walk Bombing Range

155

Start	Clay Hill, Forestry Commission car park, 2 kilometres (1¼ miles) south of Lyndhurst village centre, on the A337 Lymington road – Ordnance Survey map reference SU301062
Distance	8.5 kilometres (5¼ miles) Shorter walk options: 1) Reduce the distance by 2.75 km (1¾ miles) 2) Reduce the distance by 1.75 km (1 mile) 3) Reduce the distance by 1.25 km (¾ mile) 4) Reduce the distance by 2 km (1¼ miles) 5) Reduce the distance by 0.5 km (⅓ mile) Walk extension: A minor detour is available that adds virtually nothing to the distance
Time to allow	2 - 5¼ hours
Refreshments	The Crown Stirrup is on the route
Route	Largely along readily visible tracks, although a little 'off the beaten track' during Sections 10 and 11
Terrain	Mainly level ground
Rating	2 – moderate walking
Buggies	Not suitable
Railway station	Brockenhurst, 4.25 kilometres (2¾ miles)
Bus service	Bluestar
New Forest Tour Bus	Yes
Alternative starts	1) Parc Pale, Forestry Commission car park at Ordnance Survey map reference SU308080 2) Boltons Bench, Forestry Commission car park at Ordnance Survey map reference SU305082 (Both these start points provide access to the route at the beginning of Section 15 via a footpath leading from the B3056 Lyndhurst to Beaulieu road
'Camping in the Forest' Caravan and Campsites	1) Hollands Wood, 3 kilometres (1¾ miles) 2) Aldridge Hill, 6.25 kilometres (4 miles) 3) Matley Wood, 5.75 kilometres (3½ miles) 4) Denny Wood, 6.5 kilometres (4 miles)

SEPTEMBER
Summer slowly gives way to autumn
(Lyndhurst: Clay Hill; Parkhill, Denny, Little Holmhill, Pondhead and Park Ground Inclosures)

Fungi flourish in the driftway alongside Pondhead Inclosure

The walk
Providing an 8.5 kilometre (5¼ mile) walk in the woods south-east of Lyndhurst, this route encompasses a number of areas of ancient, unenclosed woodland and 18th and 19th century broadleaved and coniferous inclosures. Beechen Lane – now a gravelled track for much of its course, but thought to

be of medieval origin – is used, whilst quite extensive, still actively managed hazel coppices – unusual now in much of Britain – are passed along the way. A number of short-cuts are available which reduce the route by up to 4.75 kilometres (3 miles).

Featured wildlife

Badgers – autumnal diet

Badgers invariably appreciate nature's bounty, the customary late summer and autumnal glut of foodstuffs, for they depend for reasonably comfortable winter survival upon accumulating fat reserves within their bodies during these times of plenty.

A badger emerges from its sett

In September, they happily gorge on blackberries and also take wind-blown crab apples, harvests that are readily available near many of the setts. Typically damp, mild, early autumn nights also bring earthworm treats – when available, earthworms form a staple element of the badger's diet – and reliable, regular access to slugs and other soft-bodied delicacies. Fungi will also be eaten, but not as enthusiastically as acorns, whilst beechnuts and sweet chestnuts also provide appetising snacks.

More....
Badgers – boars wander widely: page 78
Badgers – rarely seen, but signs betray presence: page 205

Dor beetles – humble insects with intriguing lifestyles

The dor beetle *Geotrupes stercorarius*, often known simply as the dung beetle, is frequently seen from April until October, but is often most abundantly visible from mid-July on short-cropped, grassy woodland rides and when travelling sedately along, or across, gravel tracks. Distinctive creatures that can grow to a length of 2.5 centimetres (almost 1 inch), dor beetles possess shiny black, strongly ridged wingcases and an equally dark head and body, all with a bluish or purple iridescent sheen.

As with other related species, dor beetles are extremely fond of dung and in many New Forest woodland inclosures are frequently seen devouring, or working with, fox scats. Eggs are laid in underground nest chambers constructed within quite deep burrows – the beetles' large, spiky front legs are specially adapted for digging – but only after the chambers have been provisioned with dung set aside to satisfy the appetites of newly hatched larvae.

Dor beetles, to their cost, are widely taken by foxes, tawny owls and other predators that in the New Forest, in the relative absence of abundant small mammal populations, are no doubt pleased to supplement their diet with generous helpings of portly, well-fed beetles. In fact, fox scats used by dor beetles are often packed with dor beetle wingcases, illustrating an unusual but in some ways mutually beneficial relationship between mammal and insect.

An unfortunate dor beetle (Geotrupes stercorarius) under attack from southern wood ants

Douglas firs – tall trees with a sad story to tell

Encountered on this walk soon after entering Parkhill Inclosure during Section 3 of the route, during Section 5 and on the edge of Little Holmhill Inclosure during Section 7, the Douglas fir is one of our most imposing conifers. Introduced to Britain in 1827 by David Douglas, a botanical collector who travelled extensively in America and elsewhere, Douglas firs have been planted in the New Forest since the mid-19th century, some in small numbers as ornamentals along the edge of woodland rides, others as commercially valuable plantation trees.

Douglas firs: tall, majestic trees when allowed space in which to grow

159

The unfortunate David Douglas did not, however, see them thrive, for in 1834, at the modest age of 35, during a visit to Hawaii he reputedly fell into a pit dug to capture wild cattle and was gored or trampled to death by a bullock that had previously fallen into the trap. The incident was, however, the subject of controversy, for some at the time believed that Douglas was murdered by an escaped convict, a bullock hunter intent on robbery, who threw the body into the pit and left the bullock to take the blame.

A handsome tree when allowed sufficient space in which to develop, Douglas firs in favourable British conditions can attain a height of around 55 metres (180 feet). Mature trees have thick, deeply fissured bark in a mixture of dark grey, orange-brown and purple-brown shades; whilst cones can be readily identified by the presence of quite long, papery, three-pronged reptilian tongue-like bracts protruding from behind each scale.

Fallow deer – preparation for the rut

Fallow bucks, numbers swollen by inward migration from surrounding areas, by mid-September increasingly appear in the vicinity of traditionally used rutting stands as buck herds disperse prior to the onset of the annual rut. Antlers by then are clean of velvet, the girth of the neck has significantly increased and Adam's apples belligerently bulge.

Fallow bucks tolerate each other's company in the period leading up to the rut

Although initially content to live relatively peacefully side-by-side with other bucks, individuals increasingly advertise presence by scent marking and by fraying branches and saplings. Scrapes – shallow hollows – are also created with the aid of the forefeet and antlers, are sprinkled with urine and utilised as sources of pungent material to be spread on the body using antlers as implements and whilst wallowing.

Spiky-antlered yearling bucks engage in relatively light-hearted head-to-head jousts, but serious aggression in September is largely absent. Does and youngsters of the year, meanwhile, also increasingly gather in the vicinity of the rutting stands and, with romance in the air, all – bucks, does, youngsters and yearlings – shed at least some of their usual inhibitions and more frequently wander abroad during daylight hours.

More....
Fallow deer – always a joy to encounter: page 39
The fallow deer rut – all's fair in love and war: page 175
Bolderwood Deer Sanctuary: page 179
Fallow deer - after the rut, bucks feed well and rest: page 200

Great spotted and lesser spotted woodpeckers – a pied woodland duo

Known in the past as pied woodpeckers and barred woodpeckers, respectively, great spotted and lesser spotted woodpeckers are black and white birds with, depending upon species, age and sex, variably present and positioned patches of scarlet-red plumage. Both nest in tree holes excavated for the purpose and both lay white eggs.

Relatively common, widespread and, with a little effort, readily seen creatures; great spotted woodpeckers are blackbird-sized and have broad white shoulder patches. Occasionally in winter but more often in spring, their drumming sounds reverberate loudly through the woods as stout beaks hit solid, bark-free timber. (Each short burst of multiple blows is delivered at

A male great spotted woodpecker searches for food concealed within crevices in decaying timber

161

great speed as both a warning to competitors and a mating call.)

Lesser spotted woodpeckers are altogether more secretive and, consequently, are infrequently encountered. About the size of a house sparrow, and with white horizontal barring on back and wings, they typically remain out of sight, high in the tree tops. Reflecting the size differential, lesser spotted woodpeckers drum more quietly than their larger relatives but in more extended bursts and primarily in spring.

Calls made by both species can be heard throughout the year - sharp, clear, repeated *tchik-tchik-tchik* calls and confrontational, harsh, chattering trills from great spotted woodpeckers; and quiet *pee-pee-pee* notes from lesser spotted woodpeckers.

Yet despite their many similarities, the recent fortunes of these two birds show a marked contrast for whilst nationally, great spotted woodpecker numbers have significantly increased over the last fifteen years, or so, those of lesser spotted woodpeckers have been in sharp decline since at least the 1960s.

Marsh gentians – exquisite late summer blooms

Nationally rather scarce and, sad to say, declining long-lived perennials, marsh gentians in the New Forest are often found in small to medium sized colonies in moderately wet sections of some of the heathlands.

Lesser spotted woodpeckers are of similar size to house sparrows

The vivid colours of marsh gentian blooms contrast sharply with those of surrounding plants and grasses

During the main August and September flowering period, their quite large, bright blue trumpet-shaped blooms are carried aloft on sturdy, 10-30 centimetres (4-12 inches) high stems, but often remain unnoticed, particularly when not in flower, amongst crowding taller grasses and other vegetation. (The presence of flowers often varies between individual plants and from year to year, although those growing in more open conditions tend to bloom more than those situated in denser cover.)

Grazing by deer and commoners' stock benefits the gentians by reducing competition for space and light – the unpalatable gentians are not usually taken – whilst they also benefit from occasional light burning, but not excessive burning which can threaten their continued existence.

John Gerard, the well-known writer and herbalist, in the late 16th century noted of the marsh gentian: 'The gallant flowers hereof bee in their bravery about the end of August', and helpfully reported that the root was useful against 'pestilent diseases' and to combat the 'bitings and stingings of venomous beasts'!

Sea trout – unlikely visitors to New Forest streams

Sea trout, a migratory form of trout, spend much of their lives in the ocean, but from late summer or early autumn return to spawn in narrow, usually quite shallow streams, such as the Beaulieu River tributary running through Holmhill Passage, close to the beginning of Section 8 of this walk route. Frequently undertaken when the streams are in spate – when water levels are high enough to provide safe passage for these often quite large fish – journeys to spawn are rarely witnessed.

The Beaulieu River tributary at Holmhill Passage

Loud splashes, however, occasionally attract attention as a female trout thrashes her tail from side to side as she creates a shallow, saucer-like depression in a gravel stream bed, a 'nest' towards which she will direct her eggs whilst an attendant male simultaneously sheds sperm to fertilise them. After being covered with gravel by the female, the eggs will be left to develop in the stream until, two or three months later, they hatch. The move out to sea is made after a further year, or so, and the fish will not return to breed for several more years.

(In late autumn and early winter, sea trout may be observed, sadly marooned in relatively shallow water and with little opportunity of escape.)

Along the way

For many years fenced, though not always wholly successfully, to restrict access by deer and commoners' stock, Pondhead Inclosure, in the relative absence of grazing and, in part, as a result of continued hazel coppicing, supports a wide range and abundance of wild flowers that in late summer include brambles, creeping thistles, enchanter's nightshade, hedge woundwort, herb-robert, honeysuckle and lesser burdock.

Visits earlier in the year will also be generously rewarded with displays of bluebells, birdsfoot trefoil, common dog-violets, common cowwheat, foxgloves, greater stitchwort, lesser celandine, ramsons, wood anemones, wood-sorrel and the unlikely named wood spurge, a quite tall, patch-forming perennial with greeny-yellow flowers.

Woodland birds are also conspicuously present in Pondhead, including virtually all the common and widespread species and also some that are less expected.

Foxgloves growing in a freshly coppiced part of Pondhead Inclosure

The Route

1. Leave the south-eastern end of the Clay Hill car park – the end farthest from the main A337 road – past a set of low posts set into the ground to prevent vehicular access beyond; and continue along a wide, woodland ride beside on the left, an inclosure wood-bank and fence.

    Almost immediately, pass a wide, grassy ride on the right and eventually follow beside the wood-bank and fence as both go almost 90 degrees to the left.

    On the right at this corner is an old, in places substantial, moss-clad earthen bank, a section of the Park Pale, constructed to enclose a medieval deer park, dating back to at least the 13th century.

2. After a short distance, when adjacent to a gate on the left with a track beyond leading into Park Ground Inclosure, turn right; cross a part-wooded driftway between Park Ground and Parkhill Inclosures and reach a gate on the edge of Parkhill Inclosure.

3. Go through the gate and continue straight ahead along a wide, grassy ride. Go over a crossroads – notice the majestic Douglas firs on the right, alongside the intersecting ride at this crossroads – and just beyond, pass on the right an area of

Herb-robert – a rather delicate member of the geranium family
Herb-robert has a wonderfully long flowering season – its subtly coloured, pinky-purple petals can be seen in bloom in woods and other shaded places from April right the way through to November. It is usually found growing singly or in small clumps and can attain a height of 50 centimetres (almost 20 inches).

Country names such as redbreast, redshank and redweed reflect the fern-like foliage's tendency to redden with age, but these names are far outnumbered by others that include the words 'robin' and 'robert' – poor robin, robin's eye and bob robert are examples – derived, it's often suggested, from the plant's imagined association with Robert Goodfellow, the mischievous house and woodland goblin.

Herb-robert

woodland fenced to exclude deer and commoners' stock – this is a *Registered Seed Stand* set aside for use as a source of acorns to be sown elsewhere.

Pass at the far end of the fence-line, a wide, grassy ride on the right and again continue straight ahead.

4. At the next crossroads (where straight ahead is a grassy ride), the left turn – **the first shorter walk route** – leads along a gravel cycle track to the start of Section 12, thereby missing out the eastern and north-eastern parts of the route.

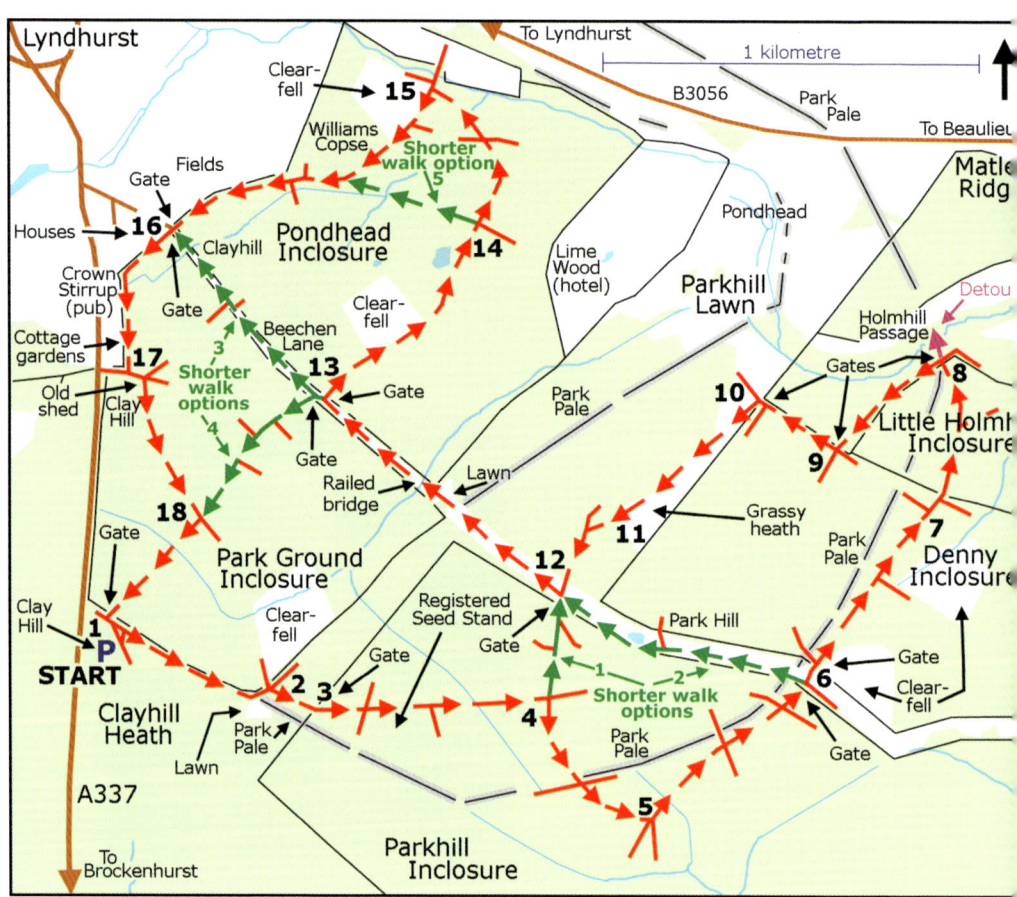

Otherwise, turn right at this crossroads; continue gradually downhill as the track bears to the left, and go straight ahead at an indistinct crossroads – where the cycle track intersects a wide, overgrown ride.

5. Turn left at the next junction – huge Douglas firs line the way here – and follow the undulating route as it eventually bears slightly to the right.
   Go straight ahead at a crossroads where another gravel cycle track goes downhill to the right, and on the left is an indistinct grassy ride.
   The Park Pale can again be seen quite close to the route along here, and also during Sections 6 and 7.
   Eventually go straight ahead again where the cycle track is intersected by a wide, grassy ride; and almost immediately pass through a pedestrian gate to emerge from Parkhill Inclosure at the end of the gravelled section of Beechen Lane.
   Beechen Lane, a historic trackway, to the left runs through a strip of unenclosed woodland that pre-dates the adjacent 18th and 19th century inclosures; whilst to the right, the original course of the old lane continues over a section of grassy heathland adjacent to a relatively recently expanded area of clear-fell.

6. **The left turn here provides access to the second shorter walk route,** which again rejoins the main route at the start of Section 12.

   Otherwise, go straight ahead and pass through a pedestrian gate into Denny Inclosure. Immediately pass a path on the left and continue along this gravel track. Eventually pass a wide, often muddy ride on the right, with to the left of it, beyond a narrow screen of conifers, an area of strongly regenerating clear-fell.

7. Just before reaching Little Holmhill Inclosure, pass a wide, grassy ride going downhill on the left and then immediately pass a number of large Douglas firs, also on the left, and another track, on the right. Continue straight ahead into the inclosure which here is unfenced.
   Eventually pass a quite wide track on the right and follow the main gravel/dirt track as it bears slightly left, downhill.

8. Leave the inclosure through a pedestrian gate.

   A short detour – continue straight ahead for a short distance to reach Holmhill Passage where a narrow stream runs through damp, primarily alder woodland containing trees with contorted, multiple trunks that show that many were in the past coppiced and/or pollarded to encourage harvests of fresh new growth.

Otherwise, immediately after leaving the inclosure, turn left to follow a quite wide path alongside the wood-bank and fence.

9. Eventually go through another gate; turn right at an adjacent crossroads; continue along a wide, grassy ride; pass another ride on the left and immediately after, go through a gate onto a section of Parkhill Lawn.

10. Immediately turn left and then bear right to cross over this fairly narrow strip of grassy heath onto which bracken has spread. Go left again and follow alongside the edge of the adjacent ancient, unenclosed woodland.

11. Very close to the end of the grassland – where straight ahead is a block of woodland – cross a narrow corridor of often wet ground, bear slightly to the right and continue gradually uphill into the ancient woodland.
    After a very short distance, turn left along a fairly indistinct, grassy track; eventually reach a gravel cycle track (Beechen Lane

Jays – subtly coloured members of the crow family

Creatures that pair for life, jays are attractively endowed with brownish-pink plumage intermixed with contrasting shades of blue, white and black. Quite large, sturdy yet surprisingly secretive birds, their loud, harsh screeches frequently provide the first indication of presence, although in autumn they can sometimes be seen in typically slow, purposeful flight, travelling in search of acorns and other foodstuffs; or else returning to their own patch of woodland to cache finds for later consumption. (Illustrating remarkable feats of memory, they have apparently been observed flying directly to cache sites to dig up stored acorns, even when snow lies thick on the ground.)

Jays relish a feast of acorns

again) opposite a gate leading into Parkhill Inclosure – from which emerges the first shorter walk route – and turn right.

12. Continue along the gravelled cycle track, eventually downhill and over a small, close-cropped lawn.

    Cross a minor stream at a railed bridge, continue uphill along Beechen Lane and eventually, when almost at the top of the hill, reach a staggered crossroads.

13. The way straight ahead here, **the third of the walk's shorter options**, rejoins the main route at Section 16 and thereby avoids Pondhead Inclosure.

    The left turn – through a gate and along a fairly wide, woodland ride – is **the fourth of the walk's shorter options**. It rejoins the main route at the start of Section 18 and thereby misses out part of Park Ground Inclosure and all of Pondhead Inclosure.

    Otherwise, go through a pedestrian gate on the right and continue along a meandering, grassy track through Pondhead Inclosure. Pass on the left, a quite large area of clear-fell beyond a narrow screen of broadleaves (where a conifer block was removed in 2006) and follow the sometimes damp track downhill to a crossroads.

Pigs and pannage – an age-old tradition

Commoners' pigs in early autumn are released into the woods to live and feed during the pannage season. Acorns, a foodstuff that if excessively eaten, particularly when green, is poisonous to ponies and cattle, are harmless to pigs and are greedily taken. The pigs, then, enjoy a free range lifestyle, satisfy their appetites, fatten for sale or slaughter and also reduce the likelihood of harm befalling other stock animals.

169

14. The left turn here, along a gravel track – the old access road to a large house originally known as Parkhill, latterly the Parkhill Hotel, and now substantially rebuilt, extended and renamed Lime Wood – provides the walk's **fifth and final shorter option**. It rejoins the main route mid-way through Section 15 and misses out the northern part of Pondhead Inclosure.

Otherwise, go straight ahead along a grassy ride past coppiced hazels in various stages of re-growth – cyclically cut, they range in age from old, overgrown examples through to freshly cut trees. Follow the ride as it eventually bears left; go straight ahead at a crossroads where the main track is intersected by a grassy path; and at the next 'T' junction, turn left along a gravel track.

The mast crop – a substantial influence on the well-being of woodland wildlife

September acorns tumble noisily from the trees, hitting the ground with a succession of dull thuds, whilst beech-mast falls with considerably less swagger. Although both are remarkably important constituents of the woodland food chain, they are, however, produced in extremely variable annual numbers, a characteristic that significantly influences the health, survival rate and subsequent breeding success of the many birds and animals that depend for food upon their availability.

Furthermore, following poor mast years, when many small mammals perish and those that survive probably breed less successfully, the health of predators of these animals – creatures such as tawny owls and buzzards – is also often badly affected and they consequently manage to raise far fewer youngsters than happens following strong mast years.

Small mammals, such as this bank vole, are an important part of the woodland food chain

More...
Magnificent oaks – so important for wildlife: page 71
Beech trees – obliging fungal hosts: page 191

170

15. Continue beside on the right, another area of 2006-cleared conifers; pass a grassy path on the left and Williams Copse on the right.

    Here a commemorative plaque inset into an aged tree stump carries the words 'Nova Foresta, New Forest, 1079-1979, nine hundred sessile oak were planted here in 1979 to commemorate the creation of the New Forest by William I in 1079. May it contribute to the sylvan pleasures of our successors'.

    Follow the gravel track as it meanders through the wood; pass after a short distance, a gravel track on the left (part of the old Parkhill drive again); pass in quick succession, an indistinct path on the right and another on the left; and continue along the track, now with fields on the right. Leave Pondhead Inclosure through a pedestrian gate and go straight across Beechen Lane.

16. Enter Park Ground Inclosure through a pedestrian gate. Continue straight ahead, almost alongside the gardens of adjacent houses on the right; and eventually follow the track round to the left beside the garden of the Crown Stirrup pub.

    Continue gradually uphill alongside, on the right, a damp meadow and then a series of cottage gardens.

17. Turn left at the next 'T' junction and then almost immediately right at a further junction, adjacent to an old shed. Continue gradually downhill along a quite open, fairly wide, grassy ride and turn right at the next crossroads.

18. Follow this equally wide ride until it eventually reaches a pedestrian and wider gate on the edge of the Clay Hill car park.

For the adventurous, for those with a good sense of direction, strong map reading skills and access to an Ordnance Survey map!
Create your own walk by combining parts of this route with elements of your choice from the following selection of connecting or conveniently located nearby routes.

From *New Forest Walks – a seasonal wildlife guide:*
March   Brockenhurst: Balmer Lawn

From *New Forest Walks – a time traveller's guide:*
Walk 8   Beaulieu Road: Black Down
Walk 11   Brockenhurst: Balmer Lawn

171

Start	Bolderwood, Forestry Commission car park, 5 kilometres (3 miles) west of Emery Down, on the Bolderwood Arboretum Ornamental Drive – Ordnance Survey map reference SU242086
Distance	8 kilometres (5 miles) Shorter walk options: 1) Reduce the distance by 5.75 km (3½ miles) 2) Reduce the distance by 4 km (2½ miles) 3) Reduce the distance by 2 km (1¼ miles)
Time to allow	2 - 5 hours
Refreshments	The New Forest Inn, Emery Down and The High Corner Inn, Linwood, are both relatively nearby
Route	Along readily visible tracks
Terrain	Undulating ground with a number of moderate gradients
Rating	3 – in places, quite strenuous walking
Buggies	Suitable for sturdy buggies if the first shorter route is used, otherwise not suitable
Railway station	Ashurst (New Forest), 10.5 kilometres (6½ miles)
Bus service	None
New Forest Tour Bus	No
Alternative starts	By the Canadian Memorial at Ordnance Survey map reference SU240093
'Camping in the Forest' Caravan and Campsites	1) Longbeech, 7.75 kilometres (4¾ miles) 2) Ocknell, 8.5 kilometres (5¼ miles)

OCTOBER
Fungi flourish and fallow deer rut
(Bolderwood Deer Sanctuary, Bolderwood Arboretum, North Oakley Inclosure, Canadian Memorial, Highland Water and Holmhill Inclosures)

A junction of tracks in Highland Water Inclosure

The walk
With its start point in the Bolderwood car park, this 8 kilometre (5 mile) primarily woodland walk passes through Bolderwood Arboretum and beside the nearby deer sanctuary, providing opportunities to enjoy close-up views of wild fallow deer. Short-cuts are available that can largely reduce the walk

to a 2.25 kilometre (1½ mile) circuit of the arboretum and deer sanctuary grounds, a prospect that might appeal to some walkers. Be aware, however, that part of this shorter route – Section 6 – is predominantly uphill.

Featured wildlife

Bats – less frequent fliers

Tucked away in hollow trees, roof spaces and other safe, sheltered places, bats by late autumn have usually hibernated, slowed down the rhythms of life in readiness for survival during the colder months when their insect prey is largely unavailable. Although occasionally tempted out by unseasonably mild weather, few will fully emerge until early spring.

Thirteen of the seventeen resident UK bat species occur locally. Common and soprano pipistrelle bats are numerous and widespread on the Crown Lands, whilst noctule, Daubenton's and Natterer's bats are also regularly encountered by enthusiasts out at night in suitable habitats with bat detectors. Nationally rare Bechstein's and barbastelle bats are notably present too, whilst within the New Forest National Park boundary, Brandt's, brown long eared, grey long eared, Leisler's, lesser horseshoe and whiskered bats have also been recorded.

A male noctule bat

Many of these bats are likely to be encountered in small numbers, flying above favoured inclosure paths and gravel tracks; within the ancient, unenclosed woodlands; and around woodland edges, but relatively large numbers often gather to hunt over insect-rich bodies of water, such as Eyeworth Pond, and in places within the woods that experience temporary gluts of insects.

Not all bats present in the New Forest during spring, summer and autumn will necessarily hibernate here, though, for at least some will probably migrate out of the area, maybe even crossing over to mainland Europe, in search of suitable wintering sites.

Birdsong in autumn – predictable in some species, but not others

Robins and wrens sing frequently in autumn as they establish and maintain territories in which to live. Less expected, though, are the occasional low volume

Robins sing throughout much of the year

songs given in October by blackbirds, chaffinches, mistle thrushes, song thrushes and a number of other birds that are not normally considered to be territorial at this time.

Why, then, do they sing? It is certainly not to express feelings of joy. Research has, however, found that the annual moult in some species, but not all autumnal songsters, causes increases in testosterone levels that could trigger low level breeding season behaviour. Other suggestions associate autumn songs with young males anxious to master the basics ahead of the breeding season; with lone migrants determined to declare presence after arriving on 'foreign' soil; and in flocking species, with efforts to maintain social status within the flock.

It is possible, of course, that the reason for song lies in a combination of some or all these factors, but for now, nobody quite knows for sure.

The fallow deer rut – all's fair in love and war

By early October, mature fallow bucks reach peak rutting condition and behaviour associated with the rut increases commensurately: vegetation is determinedly scent marked, saplings and overhanging branches are thrashed and frayed, and woodland and heathland scrapes are increasingly developed for use as wallows.

The rut usually begins in the first half of the month and lasts for around three weeks. During this period, large bucks typically take possession of a rutting stand located in the midst of woodland, at a woodland/heathland edge or else

A fallow buck groans whilst two does pointedly ignore the commotion

amongst a group of heathland trees; and during the first few of hours of daylight and again, to a lesser extent, in the evening, declare presence by groaning. Groans – loud, deep, rhythmic belch-like sounds heard only during the rut – are intended to attract does to the stand whilst simultaneously declaring powerful presence to other bucks.

Yearling and older bucks, potential competitors all, are, though, frequently attracted by the commotion. After being repeatedly chased away from the central area of the stand, yearlings usually retreat to the edge or move away completely; whilst older, stronger animals, undeterred by vocal threats and posturing, may do head-to-head battle with the incumbent in a fierce trial of strength, skill and perseverance.

Does gathered in the vicinity of the stand, meanwhile, often seem oblivious to their suitors' antics and frequently continue to feed, rest or simply stare blankly ahead as the bucks compete for their favours.

(During the period of the rut, deer watchers should take particular care not to cause disturbance by getting too close to the action. Always be aware, too, that at this time bucks (and red deer and sika stags) can be fearsome creatures that may in the excitement of the moment deliberately or accidentally blunder towards observers, with potentially dangerous consequences.)

More...
Fallow deer – always a joy to encounter: page 39
Fallow deer - preparation for the rut: page 160
Bolderwood Deer Sanctuary: page 179
Fallow deer - after the rut, bucks feed well and rest: page 200

Hedgehogs – time to hide away until spring
Sadly declining animals in many parts of Britain, hedgehogs spend late autumn and winter, the time when their favourite foodstuffs – earthworms, caterpillars, beetles, slugs and snails – are not readily accessible, in hibernation, hidden away safe from predators and the ravages of the weather, within nests constructed of grass and leaves placed amongst log piles or brushwood, concealed under garden sheds or at the base of hedgerows or bramble clumps.

Hedgehogs are rarely seen on the Crown Lands

Although occasionally abroad during spells of mild, winter weather, hedgehogs usually remain inactive until early spring, but even then, few are seen on the Crown Lands of the New Forest, probably because earthworms are relatively scarce in the typically acidic soils and possibly also as a consequence of the fairly large-scale presence of badgers, animals known to feast heartily on hedgehogs.

Winter thrushes – fieldfares and redwings, visitors from afar

Newly arrived redwings and fieldfares, autumn and winter visiting thrushes that typically breed to the east and north-east of Britain, are usually first seen in late September and early October, respectively, and significantly increase in numbers as autumn gathers pace.

Separation of the two is reasonably straightforward. Redwings are noticeably smaller than fieldfares, have prominent pale stripes above the eyes and below the cheeks, and possess relatively short tails and rusty-red flanks and underwings. Fieldfare plumage is noticeably pale grey on the rump, crown and nape, and they have particularly dark, quite long tails and white underwings.

Fieldfares enjoy haws and other berries

When in flight, the fieldfares' harsh, often quite loud, chattering calls are distinctive; whilst the redwings' thin *tseeip-tseeip-tseeip* utterances are considerably less audible. Only very rarely does either bird fully sing whilst in Britain although redwing subsong, a prolonged, quiet, twittering chorus,

A redwing searches for invertebrates

is sometimes heard, particularly as flocks settle down communally to roost; in the days immediately prior to return migration and whilst flocks rest during migratory journeys.

In autumn and early winter, both species can sometimes be seen feeding in quite large mixed flocks, frequently on holly berries or haws. Fruit will also be taken at this time and so will mistletoe, rowan and elder berries. When the availability of local foodstuffs dwindles, however, many of the birds will wander widely in search of similar fare elsewhere and may not be seen again until mid- or late winter. Then, on short-cropped grasslands they can often be watched hopping or running a few paces, stopping with head cocked to look and listen for tell-tale indications of invertebrate presence and eventually jabbing the earth to take whatever morsel has been detected. (Flocks of redwings, but rarely fieldfares, also regularly enter woodland interiors to feed amongst the leaf litter and to roost communally within the sheltered confines of hollies and other evergreens.)

By early April, most will have begun their journey back to distant breeding grounds.

Woodland and other fungi –
a harvest to be taken in moderation and with care

Visitors to New Forest woodlands from late summer until the end of November may be confronted by a sometimes bewildering array of mushrooms and toadstools, particularly in the ancient, unenclosed woods where the strongest fungal populations occur. Remarkably, more than 2,600 species have been recorded locally, making this one of the most productive fungus sites in the whole of western Europe.

Common earthballs and a variety of puffballs, some of which are edible if taken when young, are likely to be present on grasslands, woodland floors and decaying timbers, whilst other relatively readily identifiable, edible regulars include the beefsteak fungus, penny bun fungus (or cep, as it is sometimes known), parasol mushroom, oyster fungus and hedgehog fungus.

Fly agarics, the red-with-white-speckles fungi featured alongside fairies in so many children's books, may also be prominent amongst the throng. Found mainly below birch trees, fly agarics form a mutually beneficial, symbiotic relationship with the trees – fungal threads penetrate between the tree's rootlet

Fly agarics add a splash of colour to the woodland floor

178

Stinkhorns are amongst our most distinctive fungi

cells, enabling the tree to absorb nutrients via the fungus, whilst the fungus receives carbohydrates and water from the tree. Highly poisonous if taken in large quantities and reputed to induce hallucinations in smaller doses, fly agarics should never be eaten.

Other reasonably common and widespread species best left alone include virtually all the ink caps – a group so named because the caps of many, as they decay, produce a thick, dark fluid; sulphur tuft and, predictably enough, the sickener and closely related beechwood sickener. Then there's the stinkhorn, a fungus that produces a far-carrying, fetid aroma and has an unmistakable pale, hollow, cylindrical stalk and dark, slime-covered tip. Flies attracted by the stench feast on the slime and in so doing ensure the spread of this extraordinary spore-bearing substance. And to be avoided at all costs, of course, are deadly poisonous species such as the death cap, panther cap and destroying angel.

(If collecting fungi, be sure to follow the New Forest Fungi Code, as outlined in the 'Some dos and don'ts' chapter, and in particular, only pick for the table those species that can be identified with absolute certainty as safe to eat.)

More....
Fungi of the summer woods – two species: one edible, the other not: page 133
Beech trees – obliging fungal hosts: page 191

Along the way

The Bolderwood Deer Sanctuary often provides excellent opportunities to observe fallow deer in natural surroundings and at relatively close range. Although conditioned to the presence of people, the often separate groups of bucks and does are completely wild animals, free to come and go as they please but encouraged to frequent the sanctuary by the provision of spring

Fallow bucks enjoy food put out for them at the Bolderwood Deer Sanctuary

and summer foodstuffs supplied by Forestry Commission staff. Many of the deer do, however, leave these meadows at the end of summer, lured away by the prospect of acorns in the woods and the age-old urge to relocate to traditionally used rutting stands, some nearby, others much less so. But moderate numbers often remain, and rutting activity sometimes takes place in the vicinity. An impressive three-tier

Bolderwood Arboretum features a number of way-marked trails

viewing platform provides good views over the meadows, whilst interpretation boards outline the deers' annual life cycles and provide additional information. (Views over the deer sanctuary meadows are available throughout the day (and night) and there is no entry charge.)

Nearby, the Bolderwood Arboretum boasts a variety of introduced ornamental conifers planted in 1860, not long after they were first brought to Britain. (The site was originally part of the extensive grounds of Bolderwood Lodge, a master keeper's residence from at least 1732 until the house was pulled down in 1833.) Sadly, many of the original trees were lost in the great storms of 1987 and 1991, but those that remain – including Douglas fir, deodar cedar, Lawson's cyprus, western red cedar, giant redwood and black pine – continue to reach impressively skywards.

A number of easy-to-follow, way-marked trails are available within the arboretum, ranging in distance from 0.75 kilometres (½ mile) to 4.75 kilometres (3 miles). One, the Radnor Trail, passes beside the Radnor Stone, a memorial to the Earl of Radnor, Chairman of the Forestry Commission from 1952 until 1963 and Official Verderer of the New Forest from 1964 until 1966.

The Route

1. Leave the Bolderwood car park by a pedestrian gate at the south-western end, signed 'Deer viewing', and follow the gravel track as it goes round to the right. Cross the adjacent minor road; continue along the gravel track, past a sign advising 'Deer Conservation Area, Please keep dogs on leads'; and reach the Deer Sanctuary viewing platform.

2. The route from here until the start of Section 5 might seem a little tortuous, but in essence it simply follows broadly parallel to the edge of the Deer Sanctuary fields.
   Leave the viewing platform and go along a gravel track running southwards, downhill, close to the field fence-line on the right – in other words, upon reaching the platform, follow the track to the left
   Shortly after, follow the track as it goes to the right, still downhill and still close to the field boundary on the right.
   After around 100 metres, follow the track as it goes almost 90 degrees to the left. (It's always worth a short detour to the right here, to look over the fence and gate at the fields beyond. Remember, though, that access to the fields is strictly forbidden.)

3. Shortly after, turn right at a 'T' junction of gravel tracks and continue parallel to the edge of the field on the right. Almost immediately follow the track as it bears left and eventually right, ignoring on the second bend, a grassy track on the left at what effectively is a 'Y' junction.
   Go along this track as, after a short distance, it again bears left and runs parallel to the field edge on the right. Continue downhill for around 135 metres.

4. Pass a minor gravel track on the left and follow the main track as it goes to the right. Continue downhill; almost immediately go virtually straight

181

ahead at a junction, onto a wider, cycle track; and enter North Oakley Inclosure – barely discernable boundary banks are present here, but there is no gate or fence.

5. After walking a further short distance (around 65 metres) downhill, turn right at a crossroads to follow along a wide, grassy ride running through the wood towards another gravel track visible in the mid-distance. Turn right at the junction with this gravel track and go uphill along what is a further branch of the cycle network. (Short detours to the right along here, and during the first half of Section 6, will often be rewarded by sightings of deer in the adjacent fields.)

6. Leave North Oakley Inclosure through a pedestrian gate and follow the cycle track on an uphill course that has open heathland to the left, and woodland and the deer sanctuary fields to the right.

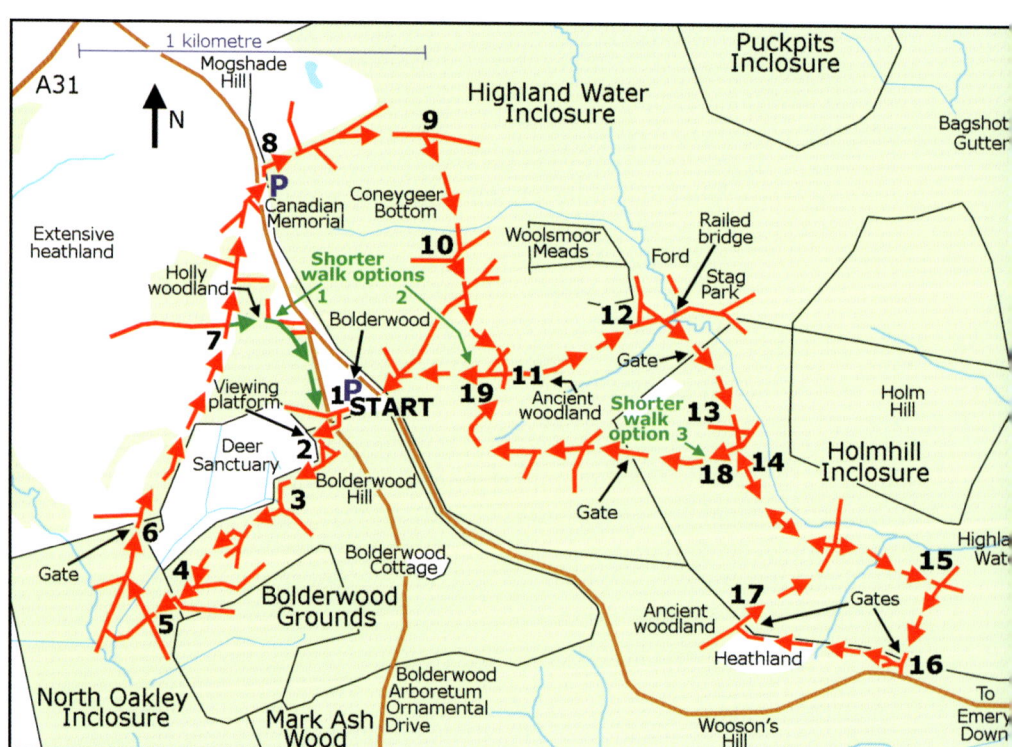

> ### Berries – stocks increasingly dwindle
>
>
> *Blackberries*
>
> Brambles continue to show occasional blossom even though many blackberries will have already ripened and been taken by birds. Sloes, too, by mid-October will have been largely gathered by people intent on making sloe gin, or else eaten by birds and animals. Rowan berries, meanwhile, will have been pilfered by blackbirds and thrushes, whilst haws will rapidly suffer the same fate. Holly berries, however, will remain on show and become increasingly conspicuous as they take on their familiar red garb, reminding that winter is not too far away.

After around 600 metres, the next crossroads, a junction of gravel tracks, provides the opportunity for an early return to the car park by restricting the walk to the area around the Bolderwood Deer Sanctuary.

7. **To take this first shorter walk route**, turn right at the crossroads and follow a minor gravel track through an area of holly woodland. Ignore three very minor paths on the left and continue round to the right, eventually parallel with, and quite close to, a minor road – the Bolderwood Arboretum Ornamental Drive. The car park is up ahead on the left.

Otherwise, continue straight ahead at the crossroads, uphill along the cycle track. Ignore minor tracks to left and right, pass beside a Forestry Commission vehicle barrier, ignore a gravel track on the right and cross a minor road close to the Canadian Memorial where a simple commemorative plaque reads: 'On this site a Cross was erected to the glory of God on 14th April 1944. Services were held here until D Day 6th June 1944 by men of the 3rd Canadian Division R.C.A.S.C.'

Turn left and almost immediately go beside a Forestry Commission vehicle barrier on the right – located on the edge of a small, gravelled car parking area – to enter Highland Water Inclosure.

8. Continue along this gravel cycle track. Pass a wide, part-gravelled ride on the left; pass a quite wide, grassy ride on the right and another on the left; and follow the cycle track as it bears round to the right, downhill.

9. Eventually turn right along the next wide, grassy ride, this one leading quite steeply downhill, and cross a narrow stream in the valley bottom – there is little threat of wet feet here as the stream runs through an underground pipe, although the track can be muddy for a short distance as it goes uphill beyond the stream.

10. Rejoin the cycle track at a staggered crossroads and continue straight ahead, uphill for a short distance. When almost at the top of the hill, at another staggered crossroads, pass a grassy ride on the left and a cycle track going uphill on the right.

    Continue downhill for a short distance before the cycle track bears round to the left, and again starts uphill. As the track goes back to the right, go straight ahead at a crossroads – the cycle track is intersected here by a grassy ride – and almost immediately reach another crossroads.

11. The right turn here offers the **second shorter route** – it avoids

European larch – one of Britain's most attractive conifers
Europe's only native deciduous conifer, the larch in autumn takes on glorious golden-yellow shades before the needles fall earthwards in a steady, colourful drizzle, blanketing all below in impressive colour. Brand new, vivid green, relatively soft needles will appear in spring, but by mid-summer will assume a darker, more foreboding hue.

Thought to have been introduced to Britain from central Europe in the early 17th century, fast-growing, relatively short-lived larches are present in the New Forest as close-planted inclosure trees and in far more attractive, smaller groupings. The extremely durable timber is said to be particularly suitable for outside use.

part of Highland Water Inclosure and all Holmhill Inclosure. Directions resume from the beginning of Section 19.

Otherwise, turn left, downhill, along a grassy ride.
Immediately on the right here is a slice of ancient, unenclosed woodland spared the axe in the 19th century when Highland Water Inclosure was created. Trees pollarded long ago are present, and so are maidens; many now with rotted, fallen trunks and boughs, their strength lost to the ravages of decay.
Note: As the land levels out along here, it might be necessary to take a short detour to the right to avoid a sometimes wet patch of ground, before rejoining the main track and continuing downhill.

12. Eventually pass a grassy track on the left and immediately turn right along a gravel cycle track – the cycle track goes virtually straight ahead - close to a railed bridge over Highland Water.
Just before the bridge is reached, go right again, along a grassy track running broadly parallel to the stream. After a short distance, go through a pedestrian gate set into a gap in the Holmhill Inclosure wood-bank and continue along an often leaf-strewn woodland path.

Dragonflies and damselflies – late fliers
Damselflies, dainty creatures that do not survive well in cold, wet weather conditions, are seen in ever-decreasing numbers as September progresses and are rarely present in October. Dragonflies, however, are typically more robust insects and, although none relish spells of bad weather, some continue intermittently to fly throughout much of October and occasionally into November. Look out, in particular, for common darters, common hawkers, migrant hawkers and ruddy darters.

Common darter

More....
Dragonflies and damselflies – colourful, primarily wetland inhabitants: page 85
Southern damselflies – jewels in the New Forest crown: page 117
The Crockford Stream: page 118
Beautiful demoiselles – incongruously exotic insects: page 137
Deadman Bottom: page 146

13. Eventually pass a wide, grassy ride going uphill to the right and immediately reach a gravel cycle track.

14. **To take the third shorter route back to the start point** (thereby missing out the final south-eastern part of the walk), turn right, continue along the cycle track as it bears further to the right and follow the route directions from the start of Section 18.

    Otherwise, follow the course of the original woodland path as it continues across the cycle route as a little used gravel track running broadly parallel to the river. Eventually continue straight ahead at a crossroads (where the walk route is intersected by a relatively narrow track).

15. On the crown of a modest-sized hill, turn right at the next crossroads where quite narrow woodland paths lie straight ahead and to the left; and continue along an overgrown, gravel track. Eventually, towards the top of a further incline, pass through a gate close to a minor road and leave the inclosure.

16. Turn right, before the road; follow an indistinct path running quite close to, and parallel with, the inclosure wood-bank; and almost immediately, at an equally indistinct 'Y' junction, take the right-hand fork to continue alongside the wood-bank – the path in many places along here barely exists, so simply stay close to the wood-bank.

    Go up a short incline before going downhill and across a small stream piped beneath the path. Cross the right-hand edge of a quite long, narrow area of heathland onto which bracken has substantially encroached; eventually follow the wood-bank as it goes quite sharply to the

> **Wild flowers – autumn blooms**
> Bell heather, cross-leaved heath and dwarf gorse blooms may still be encountered throughout much of October, complementing the rapidly fading colours of the heather and the cheery, ever-present flowers on gorse bushes. Betony, herb-robert and tormentil blooms also occasionally brighten the scene.

Tormentil (usually four petals, but occasionally five)

right; and almost immediately turn right, through a pedestrian gate giving access again to Holmhill Inclosure.

17. Continue straight ahead along an undulating, grassy path; eventually turn left at the crossroads passed mid-way through Section 14; retrace the outward part of the route back to the cycle track encountered during Section 13 and turn left along the cycle track.

18. Go uphill and eventually re-enter Highland Water Inclosure at a pedestrian gate. After a short distance, continue straight ahead at a crossroads where the cycle track is intersected by a narrow woodland path; and on the crown of the hill, pass on the left a wider, uphill path.
    Follow the cycle track, initially downhill, as it bends sharply right and then left.

19. Reach the crossroads encountered at the end of Section 10 and turn left, uphill, along a woodland track initially over-shaded by conifers. Pass an expanse of rhododendrons on the left and follow the track as it goes downhill for a short distance before resuming its uphill climb.
    When almost at the top of the hill, at the next crossroads (located in a small clearing amongst the trees), turn left along a quite narrow, gravel cycle track; follow this track for a short distance; go through a pedestrian gate and the start point car park is straight ahead.

Foxes – opportunist feeders
Foxes, often considered to be wholly carnivorous, in autumn add windfall fruits and blackberries to their already generalist diet (which also includes beetles, and earthworms unfortunate enough to have been found above ground on warm, moist nights). But perhaps most surprisingly, grass is also eaten by foxes, just as it occasionally is by pet dogs.

More...
Foxes – parents under pressure whilst raising cubs: page 100
Foxes – loud cries disturb otherwise quiet winter nights: page 206

187

For the adventurous, for those with a good sense of direction, strong map reading skills and access to an Ordnance Survey map!
Create your own walk by combining parts of this route with elements of your choice from the following selection of connecting or conveniently located nearby routes.

From *New Forest Walks – a seasonal wildlife guide:*
November   Barrow Moor, Mark Ash Wood

From *New Forest Walks – a time traveller's guide:*
Walk 4   Millyford Bridge

NOVEMBER
Woodlands ablaze with colour
(Barrow Moor, Mark Ash Wood, Wooson's Hill and Knightwood Inclosures)

A pollard beech in all its autumnal finery

The walk
November's walk just has to be in woodland for this is the month when many of our native trees take on their full autumnal splendour. And there is no better place to visit than the ancient, unenclosed woodlands of Mark Ash Wood, located 6 kilometres (3¾ miles) west of Lyndhurst. The 6 kilometre

Start	Barrow Moor, Forestry Commission car park 1.5 kilometres (1 mile) north-west of the A35, on the Bolderwood Arboretum Ornamental Drive – Ordnance Survey map reference SU254067
Distance	6 kilometres (3¾ miles) Shorter walk options: 1) Reduce the distance by 1.5 km (1 mile) 2) Reduce the distance by 2 km (1¼ miles) 3) Reduce the distance by 0.75 km (½ mile)
Time to allow	1½ - 3¾ hours
Refreshments	The New Forest Inn, Emery Down and The Swan, Swan Green, are both relatively nearby
Route	Primarily along readily visible tracks, although in places – for example, during Section 11 – a little 'off the beaten track'
Terrain	Undulating ground, but with few significant gradients
Rating	2 – moderate walking
Buggies	Not suitable
Railway station	Ashurst (New Forest), 9.5 kilometres (6 miles)
Bus service	None
New Forest Tour Bus	Yes, but only along the relatively nearby A35
Alternative starts	1) Limited roadside parking at the start of the nearby cycle route – Ordnance Survey map reference SU253068 2) Limited roadside parking at the start of Section 8 – Ordnance Survey map reference SU247072 3) Unnamed, Forestry Commission car park 300 metres south-east of the Barrow Moor car park – Ordnance Survey map reference SU257066 4) Knightwood Oak, Forestry Commission car park – Ordnance Survey map reference SU264064
'Camping in the Forest' Caravan and Campsites	1) Holmsley, 11.75 kilometres (7¼ miles) 2) Longbeech, 10 kilometres (6¼ miles) 3) Ocknell, 10.75 kilometres (6¾ miles) 4) Matley, 9 kilometres (5½ miles) 5) Hollands Wood, 10.5 kilometres (6½ miles)

Note: In recent years, both the Barrow Moor and the unnamed car park – alternative start 3 – have been closed from early November until late March. Check the Forestry Commission website for the latest information or simply be prepared to use another alternative start point.

route also takes in attractive inclosure woodland and features short-cuts that offer the potential to reduce the overall distance by 3.5 kilometres (2¼ miles).

Featured wildlife

Beech trees – obliging fungal hosts
Smooth grey bark, majestic trunk, spreading branches and breathtaking autumnal colours combine to make the beech one of our most attractive trees. Yet whilst common and widespread in the New Forest, these woodland giants suffer greatly in times of prolonged drought; are not fond of waterlogged soils; when growing in thin soils, are vulnerable to being blown over and are particularly susceptible to fungal attack, subsequent decay and consequent premature death.

Indeed, that fungi find beech attractive is most clearly evident in autumn, a time when fungal fruiting bodies add greatly to the woodland scene. Honey fungus, for example, thrives in clusters on beech and other trees, whilst the white, viscid, porcelain fungus is almost entirely restricted to beech.

Perhaps most impressive of all, though, is *Hericium erinaceus*, a fungus that in southern England is found primarily on beech. Impressive? Yes, certainly, for its pale fruiting bodies, sometimes larger than footballs, are draped in long, icicle-like spines so distinctive that they have earned for it multiple common names, such as bearded tooth, bear's head, hedgehog fungus, lion's mane, monkey head, old man's beard, satyr's beard, sheep's head and tree hedgehog. Sightings are, however, something of a rare treat, for this fungus frequently grows on boughs quite high in the canopy; and whilst the New Forest is considered to be its UK stronghold, recent surveys have found only twelve to fifteen trees with *Hericium erinaceus* fruiting bodies present.

Bearded tooth fungi (Hericium erinaceus) also have a number of other common names

But fungal attack is simply part of woodland life, unfortunate for the trees but of benefit to grey squirrels, tawny owls, jackdaws and a host of other creatures that excavate, or commandeer, cavities in decaying timber for use as roost or nest sites. And even when fallen or in the last stages of life, the trees help sustain a variety of insects, such as increasingly scarce stag beetles that during at least a part of their lives depend for survival on dead and decaying timber.

> More...
> Fungi of the summer woods – two species: one edible, the other not: page 133
> Woodland and other fungi – a harvest to be taken in moderation and with care: page 178

Bramblings – boldly marked finches that brighten the woods

Autumn and winter visiting finches from Scandinavia, Russia and other far away places, bramblings are particularly fond of beechnuts and in Mark Ash Wood can sometimes be seen searching on the ground for these fallen delicacies, often in the company of closely related chaffinches. (Although similar to chaffinches in size, shape and general appearance, the brambling's distinctive, strikingly white rump and, especially in the males, more boldly marked plumage help separate the two.)

Bramblings are autumn and winter visiting finches

Newly arrived bramblings are occasionally observed in late September, but more often in October. Numbers present each year in New Forest woodlands are significantly influenced by the strength and availability of the beechnut crop, although in the wider countryside flocks readily congregate to feed on agricultural land if a good supply of weed or other seeds is available, whilst during harsh winter weather they often turn up at garden bird feeders.

By the end of April, all will usually have started journeys back to the breeding grounds.

Common crossbills – 'errors and defects of nature'

'An error and defect of nature', 'a useless deformity'! That is how in the 18th century French naturalist Georges-Louis Leclerc, Count of Buffon, described the distinctively crossed bill tips that give crossbills their name. But he was

A female common crossbill with crossed bill tips clearly visible

very much mistaken, for the apparently grotesque twisting is, in fact, a superb adaption designed to help these stocky finches prise open conifer cones to extract their staple foodstuff: the seeds.

Common crossbills occur in many of the New Forest's coniferous inclosures and can also be seen amongst self-sown, heathland Scots pines. Numbers present vary annually, at least in part reflecting the availability, or otherwise, of foodstuffs in mainland European haunts far to the east and north-east.

Often found in small flocks, mature males are predominantly a subtle brick-red colour, whilst females are a mixture of olive-greens, greys and browns. Presence high in the trees is often betrayed by discarded pieces of pine cone tumbling to the ground, although the birds' loud, metallic *glipp-glipp-glipp* calls are distinctive and so, too, in spring and summer are the insistent sounds made by noisy fledged youngsters as they call to be fed.

Foxgloves – wisely left alone by deer and commoners' stock

Foxglove leaves in basal rosettes survive the coldest days of autumn and winter. Biennial or short-lived perennials, foxgloves grow well in many New Forest woodlands and are not significantly taken by deer or commoners' stock – the plant is very poisonous when eaten and is presumably unpalatable to animals.

From June until September, large, strikingly pink-purple – occasionally

Foxgloves grow well in many New Forest woodlands

white – tubular, bell-shaped flowers borne on tall, upright, un-branched stems, illuminate the woods, providing unexpected splashes of colour amongst the shade. Garden bumblebees and common carder bumblebees, both particularly long-tongued creatures, are regular visitors to the blooms and so are a number of other pollinating insects.

Imagined associations with foxes are, of course, fanciful and it has been suggested that the 'fox' element of the name is a corruption of 'folk', whilst local names such as fairy bells and puck fingers imply that country people of old connected fairies and goblins with foxgloves.

But despite the foxglove's poisonous nature, the plant has a long history of medicinal use as a treatment for colds, fevers, catarrh, dropsy and other ailments. Foxglove compresses were also used for ulcers, swellings and bruises, but more important was the 18th century discovery that foxglove leaves in small doses beneficially act upon the heart, which led to the eventual isolation and purification of digitoxin and digoxin for use in modern medicine as heart stimulants in the drug digitalis.

Grey squirrels – widespread New Forest colonists

Grey squirrels can hardly be confused with any other mammal, although when in summer coat, patches of reddish-brown fur could cause minor identification difficulties in the few parts of Britain where both red and grey squirrels occur. That, though, is not an issue in the New Forest, for red squirrels have long been absent following the 1930s arrival of grey squirrels, descendents of animals introduced to Britain from the USA between 1876 and 1929.

In late summer, grey squirrels feed on hazel nuts and seeds extracted from Douglas fir and other cones, leaving as evidence, empty hazel nut shells, the remains of leafy bracts that once enclosed the nuts, the central cores of stripped cones and scatterings of cone scales, all spread over the ground and over tree stumps that have been used as 'dining tables'. Then from mid- to late September, attention turns to acorns and beech-mast, some for immediate consumption, some for storage and later recovery during the cold days of winter.

Grey squirrels in winter appreciate foodstuffs cached in autumn

Grey squirrels can be aggressive creatures when confronted by even the slightest perceived threat to their well-being, angrily flicking their tails and noisily spitting out vocal warnings directed at potential predators, other squirrels and even passing deer, commoners' stock and walkers! They are, though, fully entitled to their combative attitude for they are frequently persecuted by men with guns and baited traps, and others intent on systematically destroying dens and dreys, men who object to the squirrels' occasional liking for birds' eggs and young, who dislike their much more frequent habit of stripping bark from trees to obtain access to the succulent sap layer below and who wish to protect red squirrel populations in the places where they still survive.

Buzzards, too, are enemies of the grey squirrel and so are members of the New Forest's relatively recently established but expanding population of goshawks. And of course, foxes welcome a meal of fresh squirrel meat.

Winter weather inevitably presents many challenges for the squirrels – fresh food is usually hard to find and they then largely depend for sustenance on carefully cached supplies of autumnal nuts and seeds. Yet many squirrels survive to entertain onlookers with their treetop antics.

More...
Grey squirrels – spring is in the air: page 54

Lesser redpolls – primarily autumnal and winter visitors

A rare sight in the New Forest: a male lesser redpoll in glorious breeding plumage

Relatively infrequently encountered finches, lesser redpolls feed in sometimes quite large flocks on birch and alder seeds. Although they are essentially small, streaky, greyish-brown birds, good views will reveal the mature male's pinkish-red breast, and both sexes' deeply forked tail, small black bib and the smudge of red on the forehead that gives them part of their name – the poll element derives from the Old English name for head.

Historically a somewhat reluctant visitor to garden bird feeders, lesser redpolls have developed a relatively recent liking for nijer seeds although many still seem to prefer to wander widely in search of natural food.

They are typically seen in small numbers from late September and more widely from October, and usually leave the New Forest by mid-April to return to northerly breeding grounds. But it was not always so for, following the first recorded local breeding (in 1956), they were by 1987 considered to be common and widespread breeders. Decline in the New Forest was subsequently rapid, however, mirroring a more general decline in lowland England, and by 1993 substantial decreases were suspected that were to continue unabated.

Along the way

Mark Ash Wood, despite a certain amount of storm damage suffered in relatively recent years, is considered to be one of the finest of the New Forest's ancient, unenclosed woodlands. And the name? Well, Mark Ash was not a local inhabitant, but more likely refers to an ash tree used as a boundary marker.

Although now relatively unmanaged, many of the wood's big old beech

Intertwined beech and oak trunks in Mark Ash Wood

trees are pollards, cut centuries ago at just above head height to encourage the growth of fresh new shoots safely out of reach of browsing deer and commoners' stock. Many of the hollies have also been pollarded, and some in the distant past have also been cut at ground level, in both cases to encourage the growth of shoots that would provide winter foodstuff for deer and stock.

The Route

1. Leave the Barrow Moor car park by its main entrance and turn right, along the adjacent minor road – the Bolderwood Arboretum Ornamental Drive. After a short distance, go left along a gravel cycle track (which is beside the first alternative start point).

2. Immediately ignore a grassy path on the right, continue along the cycle track and after a further short distance, as the cycle track bears left, go straight ahead at a 'Y' junction to continue along an undulating, grassy woodland ride that eventually passes over the edge of a small, tree-clad knoll.

3. Turn right at the next 'T' junction – part-way down a hill of modest gradient – and continue along a grassy, undulating ride bordered on the left by mature birch trees. Pass a quite wide, rush-strewn ride running back on the left, and after a short distance, pass another similar ride going back on the right – ride-side sweet chestnut trees along here offer the prospect of food-for-free and also striking autumnal colours.

4. Go beside an old gate post standing somewhat forlornly by a gap in the inclosure wood-bank, and leave behind Knightwood Inclosure to enter the ancient, unenclosed woodlands of Mark Ash Wood.
   Immediately bear left, continue along a path adjacent to the Anderwood Inclosure wood-bank and turn right at the next 'T' junction – an inclosure gate is on the left here.
   Follow the path as it soon bears left – ignore on the bend, a path on the right leading into the bracken. A little farther on, pass a damp hollow on the left where encroaching trees have taken hold; continue along the path as it bears slightly right; and immediately pass a grassy path on the left, running downhill into Anderwood Inclosure.
   Go straight ahead in the direction of conifers at the corner of North Oakley Inclosure, the view of which in summer is often obscured by foliage. Continue along the path as it bears right and eventually runs parallel to the

197

inclosure fence and wood-bank on the left. (Ignore minor paths on the right along here – just stay fairly close to the fence and wood-bank.)

5. Reach a crossroad of paths, with to the left, a gate and path that provide access to North Oakley Inclosure – a wooden bench here, on the edge of the conifers, offers comfortable seating for use during a refreshment or other break.

To take the first of three shorter walk routes, and miss out much of Mark Ash Wood, turn right at this crossroads and continue along a relatively indistinct, narrow path running steadily uphill for much of its course.

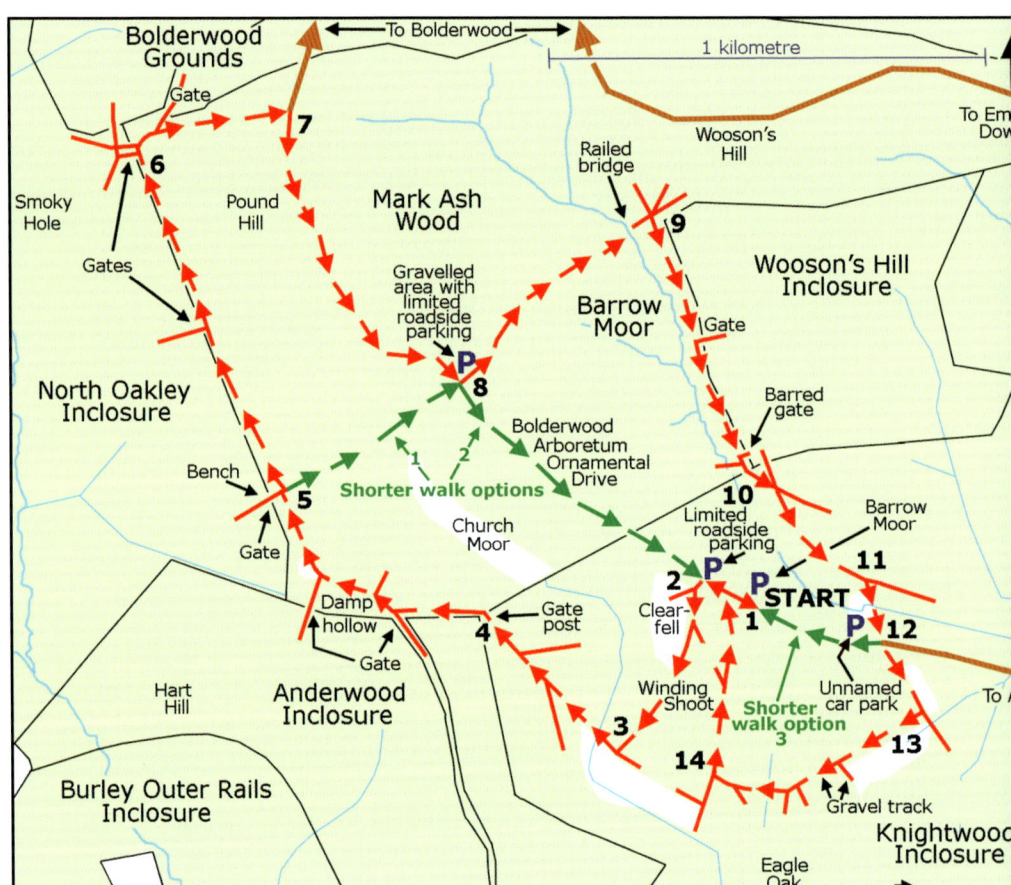

198

Eventually pass beside a low, Forestry Commission vehicle barrier, meet the minor road encountered earlier and rejoin the main route at the start of Section 8.

Otherwise, go right and immediately left to continue quite close to the inclosure fence and wood-bank on the left. Eventually pass a gate on the left giving access to North Oakley Inclosure and continue straight ahead, alongside the inclosure boundary.

6. Eventually pass a pedestrian gate on the left, with cycle track beyond; and shortly after, adjacent to a wider gate on the left, take a right-hand, quite indistinct path at a 'T' junction.

Follow this path straight ahead through the bracken, within sight

Leucobryum glaucum – the pincushion moss

Mosses sheltered from desiccating winds flourish in Mark Ash Wood, adorning tree trunks, branches, wood-banks and the woodland floor. Look out, in particular, for distinctive, closely packed *Leucobryum glaucum* 'moss balls', half-sphere shaped hummocks that are whitish when dry and dark green when wet. Also known as the pincushion moss, *Leucobryum glaucum* is often found beneath beech and oak trees.

Ferns – an oft neglected group

Scaly male fern

The Victorians perhaps appreciated ferns more than later generations, and collected specimens with such enthusiasm that wild populations were in some cases significantly reduced. A variety of ferns can still be found in the New Forest, however, including the ubiquitous bracken, a species that is rarely viewed with much affection, and widespread examples of lady fern, hard fern, broad buckler fern, male fern and common polypody. Comparative rarities, such as royal fern, beech fern and marsh fern, are also present but have relatively restricted, scattered distributions.

199

of a pedestrian gate on the left giving access to Bolderwood Grounds. Proceed uphill, initially parallel to the inclosure fence-line, and continue along a quite wide, grassy path that in places is sometimes obscured by fallen leaves.

Progressively lose sight of the inclosure boundary on the left; and when a little beyond the brow of the hill, turn right, alongside the minor road encountered earlier.

7. Follow beside the road as it eventually snakes downhill through ancient, unenclosed woodland; and shortly after it begins a gentle uphill climb, reach on the right, a low, Forestry Commission vehicle barrier and the path used during the first shorter walk route.

8. **To take the second shorter walk route**, which avoids the area around Barrow Moor, continue alongside the road to reach the car parks.

Otherwise, cross the road and walk across a small gravelled area often used for car parking; pass beside a low, Forestry Commission vehicle barrier and follow a predominantly grassy path as it meanders on an undulating course through the ancient woodland.

Eventually continue downhill; cross a narrow, alder-lined stream at a railed bridge; go uphill for a short distance, through a shallow hollow-way;

Fallow deer - after the rut, bucks feed well and rest

Mature fallow bucks in early November may still occasionally indulge in rutting behaviour, although with considerably less vigour than previously. But after feeding poorly during much of October, when finding food was of relatively low priority, and drained of energy by rutting exertions, many simply melt into the woods where they rest and regain at least some condition before gathering together into wandering 'buck herds'.

More...
Fallow deer – always a joy to encounter: page 39
Fallow deer - preparation for the rut: page 160
The fallow deer rut – all's fair in love and war: page 175
Bolderwood Deer Sanctuary: page 179

200

and turn right at the next junction of tracks.

9. Walk along an undulating woodland path beside, on the left, the Wooson's Hill Inclosure wood-bank, and on the right, sloping ground leading down to the narrow stream. Pass on the left, a pedestrian gate set into a gap in the wood-bank and eventually pass, again on the left, a gateway barred to deter access to the inclosure.

Immediately pass on the right, a path that goes downhill and across the stream; and then go through a gap in the inconspicuous wood-bank bordering Knightwood Inclosure.

10. Walk straight ahead, uphill, through an area of quite close-planted larch trees, and follow the path as it eventually bears slightly left. After a short distance, at the top of the hill, go right at a crossroads and then continue downhill along a quite wide, grassy ride.

11. At the next junction, follow another wide, grassy ride downhill to the right and cross a valley-bottom stream. (There is no bridge here, but the stream is usually fairly narrow, and in all but the wettest weather can be stepped across, or leapt, with relative ease. Indeed, in summer and early autumn the stream is often dry.)

Use a narrow path to cross a sometimes wet patch of ground beyond the stream, before climbing steeply up a short hill to reach the unnamed car park used as the walk's third alternative start point.

Woodcocks – infrequently seen woodland waders

Cryptically plumaged, long-billed, often elusive woodland waders, woodcocks feed primarily on invertebrates located by probing damp, unfrozen ground; and find shelter from the worst of the winter weather under dense rhododendron bushes and beneath low-branched holly and yew trees. Although resident numbers are swollen by autumn and winter visitors from continental Europe, sightings are often restricted to glimpses of birds that, when disturbed, lift up with a noisy clatter of wings before weaving away through the trees.

(Visitors to New Forest woodlands at dusk in spring might, however, see male woodcocks engaged in roding display flights a little above the trees or, at similar height, following the course of woodland rides.)

12. Leave the car park by its main entrance.

   To take the third shorter walk route (which misses out the final part of Knightwood Inclosure) back to the Barrow Moor car park and the nearby cycle track parking spaces, turn right and follow alongside the minor road.

   Otherwise, cross the minor road and continue along a fairly wide, grassy ride/path. After a short distance, pass beside on the left, an area in part cleared of trees where bilberry, heather, cross-leaved heath and bell heather have gained a hold.
   After a further short distance, as the ride/path goes down a moderate gradient, turn right - a little before the bottom of the hill is reached – along another grassy ride/path.

13. Go up a hill of very moderate gradient, join a gravel track at the top of this hill and continue straight ahead, now gradually downhill. Pass a fairly narrow path leading uphill to the left and almost immediately follow the main track as it bears round to the right at a junction.
   Continue gradually downhill, cross a narrow drainage channel – sometimes dry in summer – carried under the track through a pipe and pass a very overgrown, indistinct path going back on the left.

14. Almost immediately after the indistinct path, turn right at a 'T' junction and follow this gravel cycle track uphill for a short distance. As the cycle track bears left, go straight ahead at a 'Y' junction to continue along a grassy ride leading back to the minor road and the Barrow Moor car park.

For the adventurous, for those with a good sense of direction, strong map reading skills and access to an Ordnance Survey map!
Create your own walk by combining parts of this route with elements of your choice from the following selection of connecting or conveniently located nearby routes.

   From *New Forest Walks – a seasonal wildlife guide:*
   October   Bolderwood Deer Sanctuary

   From *New Forest Walks – a time traveller's guide:*
   Walk 13   Burley New Inclosure

DECEMBER
In deep mid-winter
(Fritham: Eyeworth Pond, Fritham Plain, Sloden, Woodford Bottom, High Corner Wood, Broomy Plain and Broomy Inclosure)

Mallards enjoy the sunshine and a small patch of open water on an otherwise frozen pool

The walk
Whilst days may be short, winter walks rarely fail to invigorate spirits subdued by long, dark evenings spent indoors. Starting beside picturesque, wildlife-rich Eyeworth Pond, this 13 kilometre (8 mile) heathland, woodland and

Start	Eyeworth Pond, Forestry Commission car park at Ordnance Survey map reference SU228146
Distance	13 kilometres (8 miles) Shorter walk options: 1) Reduce the distance by 1 km (0.6 miles) 2) Reduce the distance by 4.5 km (2¾ miles)
Time to allow	3¼ - 8 hours
Refreshments	The Royal Oak, Fritham, is close to the start of the walk, whilst the High Corner Inn is mid-way along the route
Route	Largely along readily visible tracks, although in places – for example during Sections 3 and 5 – a little 'off the beaten track'
Terrain	Much level ground, but with a number of uphill sections. Note: The area around Woodford Bottom, towards the end of Section 5, can be particularly wet in winter and after heavy rain, so use of strong, waterproof footwear is particularly recommended
Rating	3 – in places, quite strenuous walking
Buggies	Not suitable
Railway station	Ashurst (New Forest), 15 kilometres (9¼ miles)
Bus service	None
New Forest Tour Bus	No
Alternative starts	1) Fritham, Forestry Commission car park at Ordnance Survey map reference SU231141 2) High Corner, Forestry Commission car park at Ordnance Survey map reference SU199104 3) Milkham, Forestry Commission car park at Ordnance Survey map reference SU217102 4) Cadman's Pool, Forestry Commission car park at Ordnance Survey map reference SU230122 5) Ocknell Pond, Forestry Commission car park at Ordnance Survey map reference SU232119
'Camping in the Forest' Caravan and Campsites	1) Longbeech, 3.75 kilometres (2¼ miles) 2) Ocknell, 4.5 kilometres (2¾ miles)

wetland walk can be shortened by up to 4.5 kilometres (2¾ miles). Dense holly holms are passed along the way and so are Sloden Wood and High Corner Wood, both excellent examples of unenclosed woodland, whilst Broomy and Holly Hatch Inclosures provide further woodland variety. Extensive areas of heathland are visited and Dockens Water is twice encountered.

Featured wildlife

Badgers – rarely seen, but signs betray presence

Badgers remain at least partially active throughout the winter, but on colder nights may not emerge from their comfortable underground setts. Although relatively common and widespread in the New Forest, they rarely venture abroad before dusk, yet plentiful evidence of nocturnal activities can sometimes be seen.

Animal trails, always at their most conspicuous in winter after much of the ground-level vegetation has died back, may reveal part of the route used by a badger – the presence of tell-tale strands of coarse, black and grey hairs snagged on brambles, low branches and other obstructions will confirm Brock as the trail's originator.

Setts, often located in woodland on well-drained hill-sides, usually show signs of past or present badger occupation. Badger-made tunnel entrances, for example, are noticeably bigger than those made by rabbits, whilst spoil heaps are correspondingly large. Old, discarded bedding material – detritus from recent clear outs – may also be nearby, and clumps of dry leaves, grass or bracken dropped by over-loaded badgers carrying in new material might remain on show above ground.

Claws leave clues, too. Look out for long, distinctive impressions made in mud by badger forefeet, for scratches left by passing badgers on fallen trunks and for marks at heights of up to 1 metre (39 inches) inflicted on 'scratching trees' by animals simply stretching out their limbs or providing signs directed at other badgers.

And, of course, there are badger droppings! Usually soft and of substantial size, droppings are most often deposited in latrines – shallow

Deep 'badger' scratches on a fallen trunk

pits excavated specially for the purpose – around the periphery of setts and along territorial boundaries.

More...
Badgers – boars wander widely: page 78
Badgers – autumnal diet: page 158

Foxes – loud cries disturb otherwise quiet winter nights

Foxes can sometimes be seen hunting by day

Unlike badgers, foxes in late summer and autumn do not significantly accumulate fat reserves within their bodies to help them through the colder months. Consequently, as winter food supplies become increasingly scarce, they can sometimes be seen hunting by day, travelling purposefully along quiet woodland rides or meandering more slowly, hesitating from time-to-time to nose amongst ride-side vegetation before eventually pouncing with a typically high leap onto a small mammal, beetle or whatever other morsel has been detected.

But winter foxes do not wholly concentrate on hunting, for their annual mating season also occurs when the days are short and the nights are long and cold. Then, fox barks and other sounds are increasingly heard at dusk and after dark. Wild, often unearthly noises all, common vocalisations include hoarse *aaaaaargh*s and harsh, far-carrying, regularly repeated *wow-wow-wow*s interspersed with periods of silence. Other foxes respond with their own cries, blood-curdling screams and individual expressions of presence, affection or aggression. Direct confrontations result in much snapping and snarling.

More...
Foxes – parents under pressure whilst raising cubs: page 100
Foxes – opportunist feeders: page 187

Hen harriers – masters of the airways

Primarily autumn and winter visitors to the New Forest, hen harriers can be seen from late September until mid-April. Only very small numbers are usually present, although accurate assessment is difficult as these highly mobile creatures travel widely within the area. Heathlands and wetlands are quartered in slow, low level flight during searches for small birds and mammal prey. Night-time roosts are on the ground, out of sight amongst the heathers.

Large – up to 55 centimetres (22 inches) long and with a wingspan of up to 118 centimetres (46 inches) – relatively slender-winged predators, hen harriers breed in northern Britain and parts of Ireland, Scotland and Wales, but a liking for meals of red grouse has resulted in them becoming the UK's most intensely persecuted bird of prey, so-much-so that in recent years, in England, very few have managed to successfully raise youngsters.

Mature males are predominantly blue-grey above, mainly white below and have noticeably black wing-tips;

A beautiful male hen harrier quarters the ground as it searches for prey

whilst primarily brown females and juveniles – both known as ringtails – have conspicuous white rumps and banded tails. (Potential exists for identification confusion with superficially similar marsh harriers and Montagu's harriers, although the former is infrequently seen on the Crown Lands, whilst the latter is an extremely occasional passage migrant and even less frequent breeder.)

William Turner, a 16th century naturalist, in 1544 reported that hen harriers 'get this name among our countrymen from butchering their fowls'. Grey buzzard (an old Hampshire name), gorse harrier (from Sussex) and furze kite all, however, seem far more appropriate country names.

Mandarin ducks – exotic newcomers to ponds and streams

Originally introduced to Britain in the 18th century, mandarin ducks were initially present only as ornamental birds until small numbers escaped and spread into the countryside. They were first recorded in the New Forest in the early 1960s, but breeding was not confirmed until the early 1980s. Now, however, they are regularly seen on Eyeworth Pond and are reasonably widespread elsewhere, although on woodland streams, secretive habits often make observation difficult – presence is frequently first noticed when an agitated pair lift off from the water before weaving away through the trees and disappearing from sight. (From late winter through to early spring, in the run-up to the breeding season, pairs of mandarin ducks are also sometimes seen and heard at dusk as they fly and loudly call above favoured woodlands.)

Male plumage is a gaudy mixture of cinnamon, orange-chestnut, black and white, finished off by a dark red bill, multi-coloured crest, prominent orange side-whiskers and conspicuous 'wing-sails'. Females are a more modest mixture of greys, browns and whites.

Male and female mandarin ducks doze on the ice at Eyeworth Pond

Nuthatches – fiercely territorial woodland birds

Nuthatches are noisily present throughout the year

Noisily present throughout the year in virtually all New Forest broadleaved woodlands, nuthatches are also regular, ever-aggressive visitors to garden feeders. Dapper blue-grey and buff birds with a dash of red-brown on the flanks and relatively long, stout, pointed bills, these handsome creatures draw attention to themselves with loud, ringing, far-carrying calls and the sound of hammering as nuts wedged in bark crevices are broken open.

Year-round territories, often maintained by breeding pairs, are occupied and jealously guarded, probably to ensure access to carefully cached, often autumn gathered, stores of nuts and seeds.

Nests are typically made in natural tree holes or in old, woodpecker holes, although construction is limited to bringing in flakes of bark to form what must be a fairly uncomfortable lining, and plastering mud around the entrance to narrow its dimensions so as to exclude access by larger, potentially predatory birds.

Yet such rudimentary nesting arrangements have not prevented nuthatches from doing rather well for themselves, for since the mid-1970s, increasing numbers of nuthatch youngsters have on average been successfully raised during each nesting attempt, which has contributed to a substantial increase in the population and colonisation of parts of northern England and southern Scotland that were previously unoccupied.

Reeves' muntjac – an unwanted alien that is necessarily elusive

Reeves' muntjac – the smallest of the New Forest deer

Not much bigger than a medium sized dog and with exceptionally secretive habits, muntjac deer are infrequently seen in the New Forest although their whereabouts are sometimes revealed at dusk and after dark by barks similar to those of a terrier – muntjac are also known as the barking deer – and by small, tell-tale heaped droppings deposited on the ground.

Reliable estimates of numbers are hard to come by, although the 2013 New Forest deer census recorded the presence of fifty of these animals on the Crown Lands. However, such is the nature of these tiny beasts – originally introduced in the late 19th or early 20th century from China onto the Woburn estate in Bedfordshire – that once a significant population is established, it is virtually impossible to eradicate them from an area such as the New Forest, with consequent threats to forestry interests and native wildlife species – muntjac deer are known, for example, to cause exceptional damage to native woodland flora. They are therefore treated as a pest species that is severely culled by Forestry Commission staff, with the intention of completely eliminating them from the Crown Lands.

209

Yew – a long lived, sacred tree

Dark, sombre and rich in mystery; huge, evergreen yews are often found in churchyards, and in some cases may pre-date the earliest church on the site; but on the Crown Lands of the New Forest, they are not particularly abundant and those that are present are often considerably younger and consequently smaller than their churchyard counterparts. (The yew tree beside the Parish Church of St. Nicholas, Brockenhurst, for example, is over 1,000 years old, whilst many of the oldest Crown Land trees are considered to be a relatively modest 200-300 years old. Similarly, numerous ancient churchyard yews have girths of between nine and twelve metres (thirty - forty feet), whilst the largest yews on the Crown Lands normally measure no more than four metres – a little over thirteen feet.)

Yews do, however, grow in reasonable numbers within Sloden Old Inclosure – passed through during this walk - or at least within the section not subsumed in 1864 within the newer Sloden Inclosure. Many are likely to have originated from berries broadcast in 1771, along with holly berries, haws and sloes, to encourage the growth of trees that would provide protection for oaks grown from acorns sown at the same time.

A yew tree on the edge of Sloden Inclosure

Virtually all parts of the yew are poisonous to a greater or lesser extent, but the bark and cut, withered foliage contains more potent toxins than green foliage still on the trees, which can be taken by deer and commoners' animals, at least to moderate extent, without significant ill effect.

Male and female flowers, produced in spring, are normally borne on separate trees and the pollen is usually distributed by the wind. Berries, beautifully cup-shaped and temptingly reddish-pink and fleshy, ripen in autumn and are eaten by badgers and a range of birds, although the poisonous seed contained within is not digested, but instead is expelled to germinate elsewhere, helping, of course, to ensure survival of the species.

Along the way

Perhaps best known for its regularly present population of extravagantly plumaged mandarin ducks, Eyeworth Pond, created in the second half of the 19th century to provide water to a nearby gunpowder factory, is also home

Eyeworth Pond in the grip of winter

to a wide range of other wildlife. Canada geese, coots, mallards and moorhens, for example, fly to greet passers-by whenever there is the prospect of being fed, whilst grey herons and, primarily in winter, gadwalls and goosanders are also occasionally seen.

But it is not only water birds that are likely to be encountered, for blue tits, chaffinches, coal tits, dunnocks, great spotted woodpeckers, great tits, marsh tits, nuthatches and robins may also be present, tempted out from adjacent woodlands by nuts, seeds and bread put out for them by visitors.

(In spring and summer, resident woodland birds are joined by chiffchaffs, willow warblers, wood warblers and redstarts; whilst swallows, house martins and sand martins can sometimes be seen gracefully hawking for insects over the surface of the water. Then when the daytime birds retire to roost, at least five species of bats – common and soprano pipistrelles, Daubenton's, Natterer's and noctule bats – take to the air above the pond to begin their own searches for insect foodstuffs. Stunning white, bogbean blooms are at their best in April and May; summer water-lily displays are impressive and nearby damp ground provides ideal habitat for an array of intricately patterned mosses, bog asphodels and white, fluffy headed cottongrasses.)

Look out, too, during this walk for great grey shrikes and hen harriers – there is potential for sightings around Hallickshole Hill, Woodford Bottom, Amberslade Bottom and Ragged Boys Hill, and on Broomy, Ocknell and Fritham Plains.

The Route

Shorter walks

Option 1 – Start at the Fritham car park (Section 2) rather than at the Eyeworth Pond car park, and avoid the quite steep hill leading up from the pond.

Option 2 – Start at the High Corner car park; follow the gravelled road northwards towards the High Corner Inn; after around 100 metres, reach a 5-way junction of road, tracks and paths; and access the main route at the start of Section 7 to complete a circular walk taking in only Sections 7, 8, 9, 3, 4, 5 and 6.

The main route

1. Turn left out of the Eyeworth Pond car park and walk along the gravelled entrance road. After 200 metres, go left at a 'T' junction and continue uphill along a tarmac road bordered for much of the way by dense stands of hollies interspersed with oak and beech.
   At the top of the hill, turn right, into the Fritham car park; almost immediately ignore a gravel track on the right barred by a low, Forestry Commission vehicle barrier; and instead, follow the main gravel track as it bears left, and then continues through the car park.

2. At the far end of the Fritham car park, ignore on the right, a gravel track leading to additional car parking spaces. Pass beside a low, Forestry Commission vehicle barrier and continue straight ahead along another gravel track running across the relatively open spaces of Fritham Plain, an extensive area of heathland dotted with holly woodland, scattered hollies, oaks and occasional isolated Scots pines.
   Ignore minor tracks to left and right, and when very close to woodland directly ahead – it's around 1.5 kilometres from the Fritham car park – pass a gravel track on the right, running into Sloden Inclosure.
   Continue for a short distance, pass a minor path on the left and follow the track through a wide gap in a low, eroded wood-bank to enter Sloden Wood.

Ponies and cattle – tough days and even tougher nights

In winter, commoners' ponies can frequently be seen in woodland and within heathland gorse brakes, where they find both food and shelter. Diet during these cold months consists primarily of grass (when available), gorse and holly, but also includes twigs, bark, moss and heather.

Cattle, meanwhile, often spend the hours of daylight out in the open, and for food depend substantially on grass and, to a lesser extent, heather. By late afternoon, however, as darkness falls, many can be seen plodding stoically into woodlands in search of night-time shelter from wind, rain and cold.

The provision by commoners of supplementary foodstuffs almost invariably attracts large gatherings of animals and may prompt mini-stampedes of hungry beasts that, from a distance, hear the commoners' calls to feast.

Sloden has an impressively long, recorded history of timber production. Sloden Copse, a coppice enclosure where primarily hazel underwood was grown for fuel, wattle and other products; and oak standards provided structural materials, dates back to 1609. And whilst now unenclosed, much of the current Sloden Wood was part of an original Sloden Inclosure – sometimes referred to as Sloden Old Inclosure – set aside for timber production in around 1768. The current Sloden Inclosure is a youngster by comparison: created in 1864, it took much of the area covered by its predecessor, but not the part passed through during this walk. (The boundaries of Sloden Old and Sloden Inclosure are shown on the walk route map.)

3. Straightaway leave the gravel track by going half-right along a narrow path running parallel to the Sloden Inclosure wood-bank and fence on the right; and after 400 metres, when opposite a pronounced clearing on the left, pass on the right, pedestrian and wider gates that give access to the inclosure.

Ignore minor paths running across and alongside the clearing; when around 15 metres beyond the gates, take the right-hand fork at a relatively indistinct 'Y' junction; and follow the path between two prominent clumps of butcher's broom – a prickly, evergreen shrub - and on alongside the inclosure wood-bank and fence.

After a short distance, join a wider, grassy ride coming in from the left at the point where the inclosure fence goes off a little to the right. Continue along the path as it progressively goes away from the fence, ignore a quite

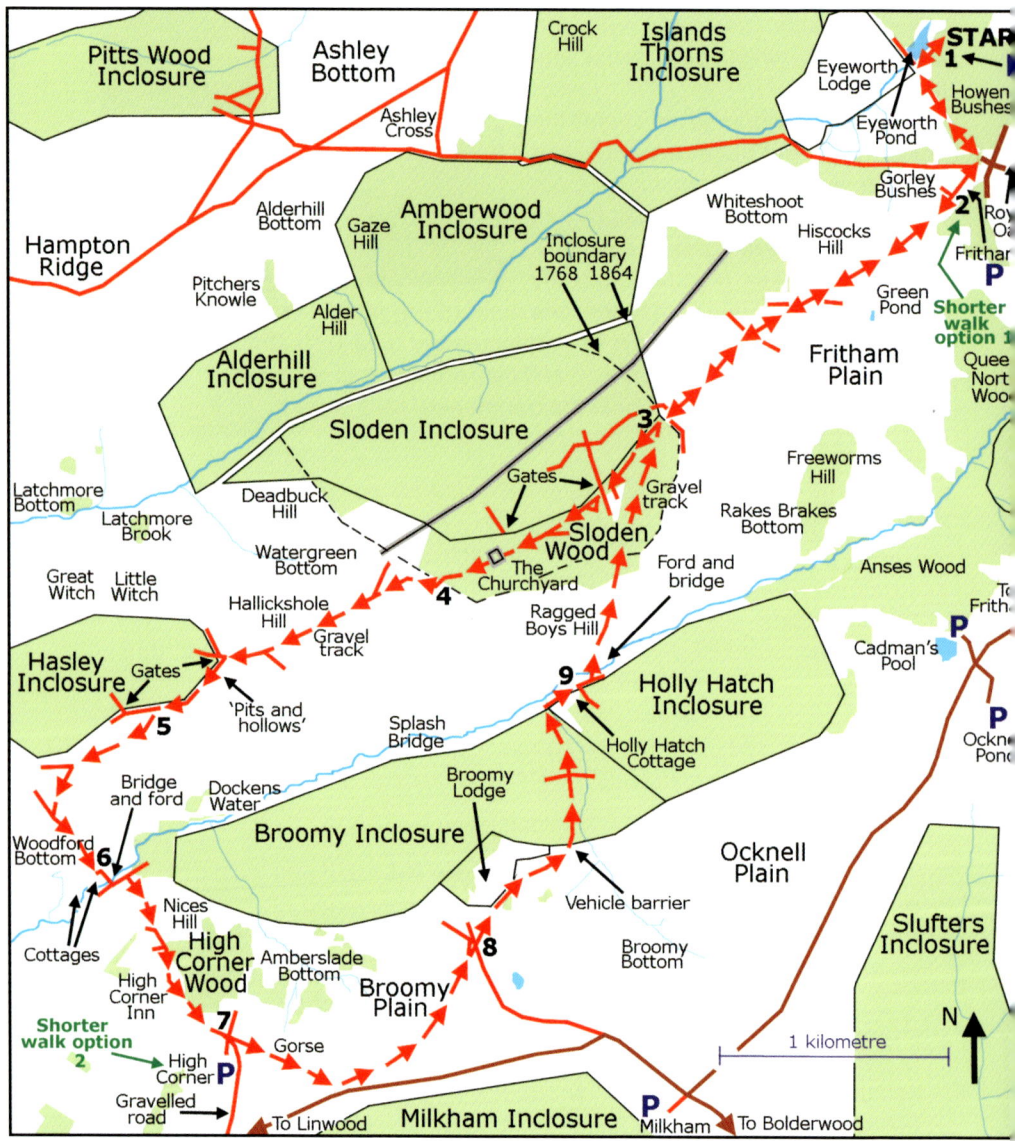

214

wide path on the left, and continue straight ahead along a broad, hill-top ridge, eventually past, on the right, another gate giving access to the inclosure.

After a further 50 metres, pass through *The Churchyard*, a quite small – around 45 x 45 metres – enclosure bounded by a now fairly inconspicuous, moss and bracken-clad earthen bank and ditch.

The site of a medieval keeper's lodge, the name reflects the once widely held but mistaken belief that William the Conqueror here destroyed a whole village, including its church, to make way for his new hunting ground. Note – when summer bracken shrouds the site, the conspicuous presence of a number of yew trees provides a strong clue to its location. (A similar lodge site, Church Place, is passed during the July 'Ashurst: Churchplace Inclosure' walk.)

Continue along the meandering path before eventually bearing left, downhill, close to the woodland edge – on the right here, beside the path, is an absolutely magnificent pollard oak with another aged oak nearby.

Woodland – winter birdlife

Woodland birds are most readily seen in winter as the all-too-often secret world of the broadleaved canopy is revealed amongst the bare branches. Tit flocks roam the woods, whilst robins, wrens and song thrushes, welcome wintry heralds of glorious dawn choruses to come, engage in joyful early morning and evening bouts of song. Old nests, too, are also prominently visible, frequently in places and numbers unimagined during spring-time walks.

Marsh tit

4. Emerge from the wood; follow the path as it continues downhill – slightly left, then right and then left again – over a hillock; before joining a quite wide, very prominent gravel track.

Go left along this track and head towards the left-hand edge of Hasley Inclosure. After around 500 metres, pass a quite wide sandy/gravel track on the left; continue up a moderate incline and immediately before reaching gates on the edge of the inclosure, turn left to follow a quite wide, grassy/sandy path running beside the inclosure wood-bank and fence.

Almost immediately on the left here is an area of disturbed ground now largely overgrown by bracken. Heywood Sumner, writing in the early 20th century, noted: '....outside the Eastern end of Hasley Inclosure, there are extensive remains of rambling pits and hollows with up-turned mounds adjoining, that may have been made for obtaining ironstone, or heathstone as it is locally called....' And separately, he again noted in respect of this site: 'The Roman potters knew the fire-resisting nature of heath-stone....and used it in their kiln constructions near by. This may have been their quarry.' (Still considered to be the site of a probable ironstone quarry, these workings are as yet undated.)

5. After around 300 metres, at a 'Y' junction, go half-left, away from the inclosure edge – this is not far from the point where broadleaved trees within the inclosure give way to a solid block of hill-top conifers.

   Continue along this grassy, undulating path as it again eventually runs broadly parallel to the inclosure edge 100 metres, or so, away; and then turn left at a junction a little before reaching the end of the inclosure away to the right. Go left at the next junction and continue across a stretch of often wet heathland until the left-most of two woodland edge cottages is reached.

6. In quick succession, go left and then right along the gravel track that gives access to the cottage, and cross Dockens Water at a railed footbridge adjacent to a ford.

   A little beyond the stream, go left along a gravelled track/road; almost immediately pass on the left, a low, Forestry Commission vehicle barrier and cycle track; and follow the track/road as it bends sharply right. For much of the way along here, continue gradually uphill; pass a thatched cottage on the right; enter High Corner Wood; pass on the right, the High Corner Inn and follow the track/road as it initially bears left, still going uphill.

   Leave the wood, emerge onto open heathland, continue uphill and after around 150 metres – where the track/road first bears right, around 100 metres before the High Corner car park – reach a 5-way junction of road, tracks and paths.

7. At this junction, leave the main gravelled track/road by taking the second turn on the left. Follow an initially quite wide but not particularly prominent, gravel/grassy track uphill through an area of gorse; and continue straight ahead, going in the direction of a minor road with beyond, the coniferous woodland of Milkham Inclosure.

   Upon reaching the road, do not cross over, but instead go left along a grassy roadside track; and follow this as it soon bears further left, away from the road. Continue around the rim of Amberslade Bottom for around 800 metres (from the road) along what quickly becomes a largely gravelled track through gorse and heather. When 200 metres from the Broomy Lodge woodland up ahead, take the right-hand fork at a 'Y' junction.

8. Almost immediately, go straight ahead at a junction and continue along a wide, gravelled cycle track. Follow this downhill, after a short distance alongside the Broomy Lodge boundary fence; and eventually, as the track bears left to enter Broomy Inclosure, pass a low, Forestry Commission vehicle barrier and sign intended for drivers – 'Holly Hatch Cottage only, no access for unauthorised vehicles'.

Continue downhill through the inclosure; go straight ahead at a crossroads; and at the far edge of the inclosure, follow the gravel track as it bears sharply right towards Holly Hatch Cottage, with here on the left, Dockens Water and its adjacent stream-side lawn.

9. Pass beside the cottage; follow the gravel track as it bears left, through a ford – there is an adjacent pedestrian bridge – and continue uphill towards Sloden Wood. Follow the gravel track through the wood and eventually retrace the early sections of the outward route to return to the Fritham and Eyeworth Pond car parks.

Siskins – brightly coloured entertainers

Small, primarily yellow-green, pleasantly vociferous finches, siskins relish conifer, alder and birch seeds, and from early autumn can be seen in large, twittering flocks, roaming the woods in agile search of nourishment. But by mid-winter, when supplies of natural foodstuffs are often exhausted, substantial numbers may turn to nuts and seeds provided at garden feeders.

In spring, some stay locally to breed and make tiny nests high in the branches of conifers, often in woodland inclosures. Others, winter visitors from northern Britain and continental Europe, undertake return migratory journeys.

For the adventurous, for those with a good sense of direction, strong map reading skills and access to an Ordnance Survey map!
Create your own walk by combining parts of this route with elements of your choice from the following selection of connecting or conveniently located nearby routes.

From *New Forest Walks – a time traveller's guide:*
Walk 2　　　　　　Bramshaw Telegraph

ACKNOWLEDGEMENTS

An enormous debt of gratitude is owed to many past chroniclers of the New Forest whose material has provided invaluable sources of reference, whilst also stimulating enthusiasm to seek out and study inhabitants of the natural world. The efforts of those who contribute wildlife records and related information to local natural history societies should also be gratefully acknowledged, and so should the writers who have used these submissions to compile outstanding accounts of Hampshire's flora and fauna. In particular, mention must be made of the publications produced or co-ordinated by Butterfly Conservation (Hampshire and Isle of White Branch), the Hampshire and Isle of Wight Wildlife Trust and Hampshire Ornithological Society.

Thanks are also due to Richard Reeves, who kindly cast his expert eyes over the chapter headed 'The New Forest: a magnificent place for wildlife' and provided invaluable comments; and to Richard Daponte of the Forestry Commission, based in Lyndhurst, who provided access to the South England Forest District Deer Management Strategy, 2014-2020 and answered numerous questions.

Similarly, thanks are due to Martin Bennett who generously provided the images of crossbill (page 193), green woodpecker (page 142), hen harrier (page 207), hobby (page 86), jay (page 168), lesser spotted woodpecker (page 162) and willow warbler (page 87); to The Hampshire Bat Group – a 'must join' organisation for anybody interested in the lives of Hampshire bats – who supplied the noctule bat image (page 174); and to Paul Brock who kindly provided the stag beetle image (page 136) and reviewed the insect images and confirmed many species identifications. (The muntjac deer (page 209) and cuckoo (page 109) images were taken by Mike Lane and sourced through the Dreamstime.com picture agency. Similarly, the goshawk image (page 65) was also sourced through the Dreamstime.com picture agency.)

And finally, I would like to thank the friends who walked these routes, checked the instructions and provided extremely helpful comments - Geoff and Mairi Aston, Steve and Margaret Boswell, Linda and Jerry Bradshaw, John and Eileen Howell, and Peter and Pat Smith.

Needless-to-say, however, any errors or omissions are entirely the responsibility of the author.

REFERENCES

Books

Birds
Clark, John, Eyre, John (editors). 1993. *Birds of Hampshire*. Winchester. Hampshire Ornithological Society.
Clark, John et al (editors). Various dates. *Hampshire Bird Reports*. Winchester: Hampshire Ornithological Society.
Cramp, Stanley et al. 1977. *Handbook of the Birds of Europe, the Middle East and North Africa – The Birds of the Western Palearctic*. Oxford: Oxford University Press.
Ferguson-Lees, J. et al. 1983. *The Shell Guide to the Birds of Britain and Ireland*. London: Michael Joseph Ltd.
Greenoak, Francesca. 1979. *All the Birds of the Air*. London: Andre Deutsch.
Holloway, Simon. 1996. *Historical Atlas of Breeding Birds in Britain and Ireland, 1875 – 1900*. London: T&A.D. Poyser Ltd.
Lever, Christopher. 1990. *The Mandarin Duck*. Princess Risborough, Bucks: Shire Publications.
Poyser. Gibbons, David Wingfield et al. 1993. *The New Atlas of Breeding Birds in Britain and Ireland: 1988-1991*. London: T&A.D. Poyser Ltd.
Svensson, Lars et al. 1999. *Collins Bird Guide*. London: HarperCollins.
Tubbs, Colin R. 1974. *The Buzzard*. Newton Abbot: David and Charles.
Wernham, Chris et al (editors). 2002. *The Migration Atlas*. London: T&A.D.

Fungi, mosses and lichens
Bon, Marcel et al. 1987. *Mushrooms and Toadstools of Britain and North-western Europe*. London: Hodder & Stoughton.
Dickson, Gordon, Leonard, Ann et al (editors). 1996. *Fungi of the New Forest: a Mycota*. Manchester : British Mycological Society.
Dobson, Frank S. 2005. *Lichens: An Illustrated Guide to the British and Irish Species*. Slough, Berks: The Richmond Publishing Co Ltd. Jahns, Hans Martin. 1980. *Ferns, Mosses and Lichens of Britain and Northern and Central Europe*. London: William Collins & Co Ltd.
Pacioni, Giovanni. 1985. *MacDonald Encyclopedia of Mushrooms and Toadstools*. London: MacDonald & Co (Publishers) Ltd.

Insects
Asher, Jim et al. 2001. *Millennium Atlas of Butterflies in Britain and Ireland.* Oxford: Oxford University Press.
Brock, Paul D. 2011. *Photographic guide to Insects of the New Forest and surrounding area.* Newbury, Berks: Pisces Publications.
Edwards, Mike and Jenner, Martin. 2009. *Field Guide to the Bumblebees of Great Britain and Ireland.* Eastbourne, East Sussex: Ocelli.
Gibbons, Bob. 1986. *Dragonflies and Damselflies of Britain and Northern Europe.* London: Hamlyn Publishing Group Ltd.
Oates, Matthew, Taverner, John, Green, David et al. 2000. *Butterflies of Hampshire.* Newbury, Berks: Pisces Publications.
Powell, Dan. 1999. *Guide to the Dragonflies of Great Britain.* Chelmsford, Essex: Arlequin Press.
Prŷs-Jones, Oliver E. and Corbet, Sarah A. 2014. *Bumblebees.* Exeter: Pelagic Publishing.
Taverner, John, Cham, Steve, Hold, Alan et al. 2004. *Dragonflies of Hampshire.* Newbury, Berks: Pisces Publications.
Thomas, J.A. 1989. *Butterflies of the British Isles.* London: Hamlyn Publishing Group Ltd.

Mammals
Burrows, Roger. 1968. *Wild Fox.* Newton Abbot: David and Charles.
Fawcett, John K. 1997. *Roe Deer.* London and Fordingbridge, Hants: The Mammal Society and The British Deer Society.
Fawcett, John K. 2003. *New Forest Roe Deer.* London and Southampton: People's Trust for Endangered Species and Mammals Trust UK.
Holm, Jessica. 1987. *Squirrels.* London: Whittet Books Ltd.
Langbein, Jochen and Chapman, Norma. 2003. *Fallow Deer.* London and Fordingbridge, Hants: The Mammal Society and The British Deer Society.
Macdonald, David and Barrett, Priscilla. 1993. *Mammals of Britain and Europe.* London: Harper Collins Publishers.
Neal, Ernest. 1986. *Natural History of Badgers.* Beckenham, Kent: Croom Helm Ltd.
Page, Andy. 2014. *South England Forest District Deer Management Strategy, 2014-2020.* Lyndhurst: Forestry Commission.
Prior, Richard. 1987. *Deer Watch: Watching Wild Deer in Britain.* Newton Abbot: David and Charles.
Putman, Rory. 2000. *Sika Deer.* London and Fordingbridge, Hants: The Mammal Society and The British Deer Society.
Russell, Valerie. 1976. *New Forest Ponies.* Newton Abbot: David and Charles.

Taylor Page, F. J. (editor). 1971. *Field Guide to British Deer.* Oxford: Blackwell Scientific Publications.
Yalden, Derek. 1999. *History of British Mammals.* London: T&A.D. Poyser Ltd.

New Forest

Edlin, H.L. (editor). 1951. *New Forest – Forestry Commission Guide.* London: HMSO.
Heathcote, Terry. 1990. *A Wild Heritage: The History and Nature of the New Forest.* Southampton: Ensign Publications.
Heathcote, Terry. 1997. *Discovering the New Forest.* Tiverton, Devon: Halsgrove.
Lascelles, Gerald. 1998. *Thirty-Five Years in the New Forest.* Lyndhurst: New Forest Research and Publications Trust.
Pasmore, Anthony (editor). 1993. *New Forest Explosives: An Account of the Schultze Gunpowder Company of Eyeworth and the Armaments Research Department, Millersford.* Winchester: Hampshire Field Club and Archaeological Society.
Read, Chris. Proceedings of the Hampshire Field Club and Archaeological Society, Volume 54: *Ancient New Forest trees.* Cheltenham: Archetype IT Ltd.
Smith, Nicola. Proceedings of the Hampshire Field Club and Archaeological Society, Volume 54: *The Earthwork Remains of Enclosure in the New Forest.* Gloucester: Alan Sutton Publishing Ltd.
Stagg, David. Proceedings of the Hampshire Field Club and Archaeological Society, Volumes 45, 46 and 48: *Silvicultural Inclosure in the New Forest to 1977.* Gloucester: Alan Sutton Publishing Ltd.
Sumner, Heywood. 1923. *A Guide to the New Forest.* Ringwood: Brown and Son.
Tubbs, Colin R. 1968. *The New Forest: An Ecological History.* Newton Abbot: David and Charles.
Tubbs, Colin R. 1986. *The New Forest: A Natural History.* London: William Collins Sons & Co.
Wise, John R. 1863. *The New Forest: its History and its Scenery.* London: Smith, Elder and Co.

Reptiles

Appleby, Leonard G. 1975. *British Snakes.* London: John Baker (Publishers) Ltd.
Frazer, Deryk. 1983. *New Naturalist, Reptiles and Amphibians in Britain.* London: William Collins Sons & Co.

Trees

Edlin, Herbert L. 1978. *The Tree Key.* London: Frederick Warne (Publishers) Ltd.
Miles, Archie.1983. *Field Guide to the Trees and Shrubs of Britain.* London: Readers Digest Association Ltd.

Miles, Archie. 1999. *Silva: The Tree in Britain.* London: Ebury Press.
Miles, Archie. 2006. *The Trees that made Britain.* London: BBC Books.
Milner, J. Edward. 1992. *The Tree Book.* London: Collins and Brown Ltd.
Selby, Prideaux John. 1842. B*ritish Forest Trees.* London: John Van Voorst.

Wild flowers

Akeroyd, John. 2001. T*he Encyclopedia of British Wild Flowers.* Bath: Parragon.
Attenborough, David. 1995. *The Private Life of Plants.* London: BBC Books.
Brewis, Anne, Bowman, Paul and Rose, Francis. 1996. *The Flora of Hampshire.* Colchester: Harley Books.
Culpeper, Nicholas. 1826 edition. (Original edition: 1652/1653). *Culpeper's Complete Herbal and English Physician.* Manchester: J. Gleave and Son.
Fitter, Alastair. 1987. *Wild Flowers of Britain and Northern Europe.* London: William Collins Sons & Co Ltd.
Fitter, Richard, Fitter, Alastair and Blamey, Marjorie. 1974. *The Wild Flowers of Britain and Northern Europe.* London: William Collins Sons & Co Ltd.
Gerard, John. 1994 edition. (Original edition: 1597). *Gerard's Herball.* London: Senate.
Grigson, Geoffrey. 1996 edition. *The Englishman's Flora.* Oxford: Helicon Publishing Ltd.
Lang, David. 2004. *Britain's Orchids.* Old Basing: Hants: Wild Guides.
Mabey, Richard. 1996. **Flora Britannica**. London: Sinclair-Stevenson.
Press, J.R. et al. 1994. *Field Guide to the Wild Flowers of Britain.* London: Readers Digest Association.
Sutton, David. 1989. *The Complete Guide to Wild Flowers.* London: Grisewood & Dempsey Ltd.

General texts

Clare, John. 1984. *The Oxford Authors – John Clare.* Oxford: Oxford University Press.
Holmes, Nigel et al. 1980. *The Living Countryside.* London: Readers Digest Association.
White, Gilbert. 1978. *The Natural History of Selborne.* Gilbert White Museum Edition. London: Book Club Associates.

Maps

Isaac Taylor's Map of Hampshire and the Isle of Wight, 1759

Web resources

Arkive: dor beetles – www.arkive.org/dor-beetle/geotrupes-stercorarius/#text=Facts

BirdGuides: Cuckoo migration –
www.birdguides.com/webzine/article.asp?a=3613
BBC Nature: autumn birdsong –
www.bbc.co.uk/blogs/natureuk/2011/10/autumn-bird-migration-news-ful.shtml
British Dragonfly Society – www.british-dragonflies.org.uk/home
British Medical Journal: Adder bites in Britain –
www.ncbi.nlm.nih.gov/pmc/articles/PMC1687390/
British Trust for Ornithology – www.bto.org (various references)
Bumblebee Conservation Trust –
  www.bumblebeeconservation.org/about-bees/lifecycle/
Bumblebee.org – www.bumblebee.org/bodyLegs.htm
Butterflies of Britain and Europe –
  www.learnaboutbutterflies.com/index.htm
Emonocot (a biodiversity web-resource for monocot plants) – http://e-monocot.org/gladiolus-illyricus-new-forest-one-britain-s-most-enigmatic-and-beautiful-monocots
Forestry Commission:
  Conservation Designations – www.forestry.gov.uk/forestry/INFD-6A5KW3
  New Forest District Deer Management Plan –
    www.forestry.gov.uk/pdf/new-forest-deer-plan-2005-2015.pdf/$FILE/new-forest-deer-plan-2005-2015.pdf
  New Forest Inclosures – www.forestry.gov.uk
Froglife – www.froglife.org/
Hampshire Biodiversity Action Plan –
  www.hampshirebiodiversity.org.uk/pdf/PublishedPlans/Southern DamselflyjjDTP.pdf
Kew Royal Botanic Gardens – www.kew.org/ (various references)
Natural History Museum: the hornet – www.nhm.ac.uk/nature-online/species-of-the-day/collections/our-collections/vespa-crabro/
New Forest National Park Authority – New Forest bats
  www.newforestnpa.gov.uk/looking-after/wildlife/bats/new-forest-bat-species
RSPB – www.rspb.org.uk/ (various references)
UK Biodiversity Action Plan –
  http://webarchive.nationalarchives.gov.uk/20110303145213/
  http://ukbap.org.uk/default.aspx
Wikipedia: dor beetles –
  http://en.wikipedia.org/wiki/Geotrupes_stercorarius

INDEX

Adder **103, 145**
Alderhill Inclosure 37
Amberwood Inclosure 37
Armaments Research Department, Millersford 152
Ashley Lodge 50
Ashley Walk 139
Azure damselfly 119

Badger **78, 158, 205**
Balmer Lawn 69
Bank vole 170
Barrow Moor 189
Bartley Cricket 53
Bartley Water **60**
Bats **174,** 211
Beautiful demoiselle damselfly 61, 119, **137,** 146
Beaulieu Airfield 122
Beaulieu Heath 113
Beaulieu River 89
Beaulieu Road 83
Beech **191**
Beech fern 199
Beechen Lane 167
Beechwood sickener 179
Beefsteak fungus 178
Bell heather **143,** 186
Berries **183**
Betony 186
Birdsfoot trefoil 164
Birdsong
   Autumn **174**
   Mid-winter **46**

Bishop's Dyke 90
Black Down 83
Blackberries 183
Blackthorn **77,** 183
Blue-tailed damselfly 119
Bluebell 90, 164
Blue tit 211
Bog asphodel **141, 211**
Bog myrtle **114**
Bog orchid **122**
Bogbean 211
Bolderwood Arboretum 173, **180**
Bolderwood Deer Sanctuary 173, **179**
Bomb craters 77
Bracken 199
Brambling **192**
Brimstone 61, 96, 132
Broad buckler fern 199
Broad-bodied chaser 85, 119
Broomy Inclosure 203
Broomy Plain 203
Bugle 90
Bumblebees **74**
Busketts Lawn 53
Busketts Lawn Inclosure 53
Butcher's-broom **38**
Butterflies **61, 96, 132**
Buzzard **38,** 76, 170

Canada goose 211
Canadian Memorial 173, 183
Castleman's Corkscrew 97
Cep 178
Chaffinch 211

225

Chicken of the woods 133
Chiffchaff **80**, 211
Church Place, Ashurst 138
Churchplace Inclosure 127
Churchyard, Sloden 215
Clay Hill 157
Coal tit 75, 211
Cockchafer beetle 136
Cockley Hill 37
Comma 61, 96, 132
Common blue damselfly 119
Common cow-wheat 164
Common crossbill 75, **192**
Common darter 85, 146, 185
Common dodder **123**
Common dog-violet 79, 164
Common earthball 178
Common hawker 85, 185
Common lizard **103**, **145**
Common polypody 199
Coniferous woodland **75**
Coot 211
Crab apple **141**
Crockford Bridge 113
Crockford Stream 113
Cross-leaved heath **143**, 186
Cuckoo **109**
Cunninger Bottom 139
Curlew 89, **94**, **99**, 146

Dartford warbler **84**, 89, 146
Deadman Bottom 139
Death cap **179**
Deerleap Inclosure 127
Denny Inclosure 157
Destroying angel 179
Dor beetle **158**
Douglas fir **159**
Downy emerald 119

Dragonflies and damselflies 85, 118, **185**
Dunnock 211
Dwarf gorse **154**, 186

Emperor dragonfly 119, 146
Enchanter's nightshade 164
Eyeworth Pond 203, **210**

Fallow deer **39**, 43, 146, **160**, 175, **200**
Ferns **199**
Ferny Crofts 83
Fieldfare **177**
Firecrest 75
Fly agaric **178**
Four-spotted chaser 146
Fox **100**, **187**, **206**
Fox Hill 53
Foxglove 164, **193**
Frame Heath Inclosure 69
Fritham Plain 203
Frohawk, F.W. 130
Fungi **133**, **178**
Furzy Lawn Inclosure 53

Gadwall 211
Ganoderma adspersum **133**
Gatekeeper 132
Germander speedwell 90
Godshill Cricket 37
Godshill Inclosure 139
Goldcrest 75
Golden-ringed dragonfly 118, 146
Goldfinch **70**
Goosander 211
Gorse **64**, 186
Goshawk **65**, 76
Grass snake **103**, **145**
Grayling **151**

Great grey shrike **41**, 42, 89, 105, 212
Great spotted woodpecker **161**, 211
Great tit 211
Greater stitchwort 90, 164
Green woodpecker **142**
Greenfinch 89
Grey heron 60, 211
Grey squirrel **54**, 191, **194**
Grey wagtail 61, **93**

Hale Purlieu 139, 146
Hampton Ridge 37
Hard fern 199
Hawfinch 42
Haws 183
Hazel catkins **47**
Heath spotted orchid **115**
Heather **143**, 186
Hedge woundwort 164
Hedgehog **176**
Hedgehog fungus 178
Hen harrier 42, 89, 105, **207**, 212
Herb-robert 90, 164, **165**, 186
Hericium erinaceus **191**
High Corner Wood 203
Highland Water Inclosure 173
Hobby 76, **85**, 146
Holly **54**, 183
Holly blue 96, 132
Holmhill Inclosure 173
Holmsley Bog 97
Holmsley Inclosure 97
Honey fungus 191
Honeysuckle **70**, 164
Hornet **128**
House martin 211

Ink cap 179
Ironstone quarry (probable) 215
Ivy **55**

Jackdaw 191
Jay **168**

Keeled skimmer 90, 119, 146
Kestrel 76, **144**
Kingfisher **56**, 61
Knightwood Inclosure 189

Lady fern 199
Lapwing 89, **94**
Larch **184**
Large skipper 132
Large red damselfly 85, 119
Large white 132
Lesser burdock 164
Lesser butterfly-orchid **123**
Lesser celandine 79, 90, 164
Lesser redpoll **195**
Lesser spotted woodpecker **161**
Leucobryum glaucum **199**
Lichens **57**
Linnet 89, **115**
Little Holmhill Inclosure 157
Long-eared owl 75

Male fern 199
Mallard 60, 211
Mandarin duck 60, **207**
Mark Ash Wood 189, **196**
Marsh fern 199
Marsh gentian **162**
Marsh tit 211
Mast crop **170**
May bug 136
Meadow brown 132
Meadow pipit 89, **109**
Merlin 105
Migrant hawker 85, 185
Millersford Bottom 139
Millersford Plantation 139

Moorhen 211
Moss **199**

New Copse Inclosure 69
New Forest
 Cattle **213**
 Designations 13
 Landscapes 9
 Management 10
 Ponies **213**
Nightjar 89, **110**, 146
North Oakley Inclosure 173
Nuthatch **208**, 211

Oak **71**
Oyster fungus 178

Painted lady 96, 132
Pannage **169**
Panther cap 179
Parasol mushroom 178
Park Ground Inclosure 157
Park Pale 165, 167
Parkhill Inclosure 69, 157
Peacock 61, 96, 132
Pearl-bordered fritillary **101**
Penny bun fungus 178
Peregrine falcon 105
Perrywood Haseley Inclosure 69
Pig Bush 83
Pignal Inclosure Inclosure 69
Pignalhill Inclosure 69
Pigs **169**
Pitts Wood Inclosure 37
Pondhead Inclosure 157, **164**
Primrose 79
Public transport 32
Puffball 178

Rabbit **44**
Ramsons 90, 164
Raven 76, **152**
Red admiral 96, 132
Red deer **102**
Redshank 89, **94**
Redstart **116**, 211
Redwing **177**
Reeves' muntjac **209**
Reptiles **103**, **145**
Ringlet 132
Ringwood Manor 48
Robin **120**, 211
Roe deer **58**, **128**, 146
Rowan **147**, 183
Rowbarrow 83
Royal fern 199
Ruddy darter 85, 185
Rushpole Wood 53

Sand lizard **103**, **145**
Sand martin 211
Scots pine **106**
Sea trout **163**
Shappen Hill 97
Shatterford 83
Shipton Holms 113, 124
Sickener fungus 179
Sika deer **73**
Silver birch **86**
Silver-studded blue **117**
Silver-washed fritillary **129**
Siskin 75, **217**
Skylark 48, 89
Sloden 203, 213
Sloden Inclosure 37, 213
Sloes 183
Slow worm **103**, **145**
Small heath butterfly 132
Small skipper 132

228

Small pearl-bordered fritillary **101**
Small red damselfly 90
Small tortoiseshell 61, 96
Small white butterfly 132
Smooth snake **103, 145**
Snipe **59**, 89, **94**, 146
Southern damselfly **117**, 119
Southern hawker 146
Southern wood ant **42**
Sparrowhawk 75
Speckled wood **96**
Stag beetle **136**, 192
Standing Hat 69
Stinkhorn **179**
Stonechat 89, **104**, 146
Sundews **146**
Swallow 211
Sweet chestnut **130**

Tawny owl **66**, 75, 170, 191
Tit flocks **124**
Tormentil 186
Turf Hill 97

Volunteer Rifle Range 45

Walks
  Alternative routes 26
  Buggies 27
  Footpath and car park closures 29
  Footwear 27
  Maps 27
Water-lily 211
Wheatear **95**
White admiral 132
Wild flowers **79, 90, 186**
Wild strawberry 90
Wild gladiolus **131**
Wildlife 12
  Habitats 10

Safeguards 33
Watching
  Binoculars 20
  Clothing 20
  Field guides 21
  Fieldcraft techniques 23
  Observation skills 23
  Taking notes 21
  Timing of visits 22
  Weather conditions 21
Willow **74**
Willow warbler **87**, 211
Wood anemone 90, 164
Wood sage **135**
Wood spurge 164
Wood warbler **87**, 211
Wood-sorrel **88**, 90, 164
Woodcock **201**
Woodfidley 83
Woodford Bottom 203
Woodland birds
  Spring **108**
  Winter **215**
Woodlark **48**
Wooson's Hill Inclosure 189
Wren **131**

Yew **210**
Yew Tree Heath 83
  Second World War anti-aircraft
    battery command post 95

229

NOTE

In this book's companion publication, *New Forest Walks – a time traveller's guide*, walk number cross-references refer to the following 'seasonal wildlife walks':

Walk 1	January
Walk 2	August
Walk 3	December
Walk 4	February
Walk 5	July
Walk 6	April
Walk 7	June
Walk 8	March
Walk 9	September
Walk 10	November
Walk 11	October
Walk 12	May

Also from Sigma Leisure:

New Forest Walks
Discovering The Past – A Time Travellers Guide
Andrew Walmsley
Explore the New Forest with this series of 16 walks through ancient landscapes where long-forgotten bumps, hollows and moss-clad earthen banks have stories to tell of Bronze and Iron Age peoples, Romans, Normans and others who lived, worked and hunted here. Illustrated throughout with colour photographs.
£12.99

Family Walks in Hampshire
in and around The Meon Valley
Louis Murray
The river Meon is one of Hampshire's quintessential chalk streams. It rises from natural springs in the South Downs to the south of the village of East Meon. This book contains the details of 20 walks in the Meon river valley area in southern Hampshire. The walks are suitable for novices, casual walkers, family groups, and experienced ramblers.
£8.99

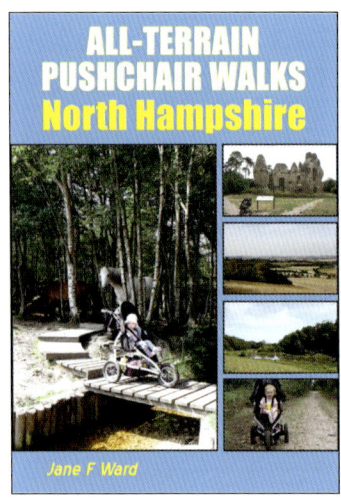

All-Terrain Pushchair Walks: North Hampshire
Jane F Ward

30 carefully selected all-terrain buggy walks in beautiful North Hampshire. From strolls through ancient forests, heathland rambles to spectacular uplands romps. Whether you're walking to keep fit or to enjoy the great outdoors.

£8.99

South West England Nature Walks
from train stations
Graeme Down

Get away from it all and find nature using the train! Although the countryside is easily accessible by car, it's far more relaxing to combine the beauty of the countryside with the less stressful mode of train travel. Twenty-four circular walks are described, starting and finishing at stations across the south-west, keeping off road as much as possible, and taking the walker through landscapes from fens to farmland and coastal surf to chalk downland. There are two walks for each month of the year, timed to give maximum chance of spotting the wildlife on offer. Each route is clearly described and accompanied by a map. Along the way, hints are given to help the reader identify some of the wildlife they may be able to find.

£8.99

Dorset Accessible Walks
25 accessible walks in the beautiful county of Dorset
Marie Houlden

All of the walks are stile and obstacle free, with consideration given to those in wheelchairs. With walks that start from only a mile and that cover a mixture of terrain and environments, there really is something for everyone. There are even a couple of more strenuous walks for those with an all-terrain pushchair and a passion for a physical challenge!

£8.99

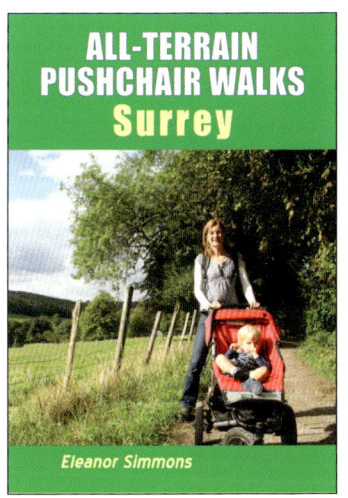

All-Terrain Pushchair Walks Surrey
Eleanor Simmons

A collection of 30 varied and multi-terrain pushchair friendly walks in the beautiful Surrey countryside. Enjoy gentle woodland and riverside strolls and more strenuous hikes in the Surrey hills. Each of the routes has a map, directions and essential information for fun and easy walking with babies and young children.

£8.99

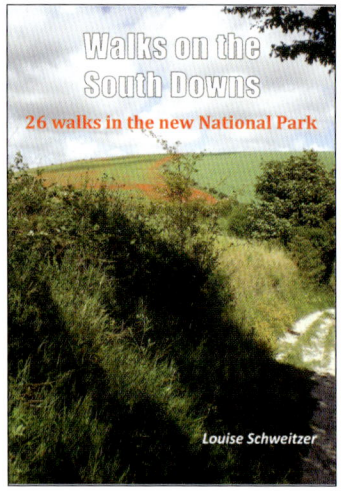

Walks On The South Downs
26 Walks in The New National Park
Louise Schweitzer

Twenty five circular trails range from five to ten miles long around some of the most unspoilt and spectacular scenery in England on waymarked public footpaths, bridleways, old coach roads and an occasional tarmac lane. Most routes feature a particular landmark, viewpoint, monument or preserved antiquity, touching the more familiar long distance trails in passing, but creating new viewpoints for the present from some perspectives of the past.
£8.99

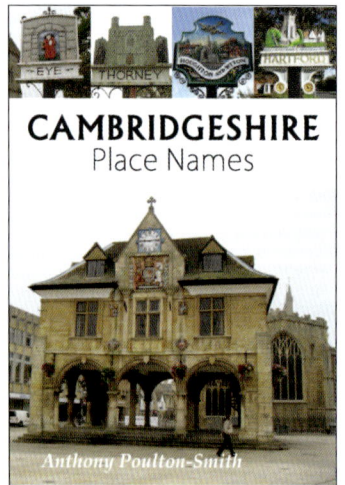

Cambridgeshire Place Names
Anthony Paulton-Smith

Ever wondered why our towns and villages are named as they are? Who named them and why? Towns, villages, districts, hills, streams, woods, farms, fields, streets and even pubs are examined and explained. The definitions are supported by anecdotal evidence, bring to life the individuals and events which have influenced the places and the way these names have developed.This is not simply a dictionary but a history and will prove invaluable not only for those who live and work in the county but also visitors and tourists, historians and former inhabitants, indeed anyone with an interest in Cambridgeshire.
£8.99

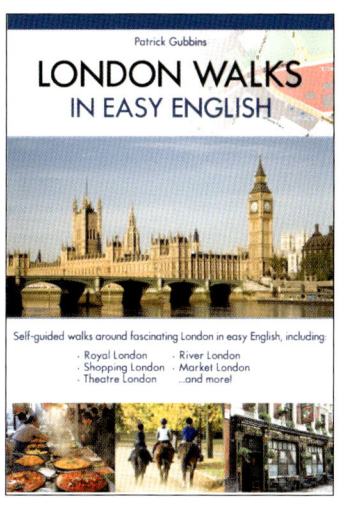

London Walks in Easy English
Patrick Gubbins

Forget the boring "walk books" that take you down quiet streets where nothing happens. *London Walks in Easy English* knows where the busy, exciting places in the capital are, and makes sure you see London life with all its colour, tradition, food, views, art, beautiful buildings and, most importantly, its sense of fun. What other book of walks takes you inside the classrooms of London University, into courtrooms to see real trials in progress, into shops to try exotic food, and to the big attractions but also to many other fascinating places that even Londoners don't know?
£9.99

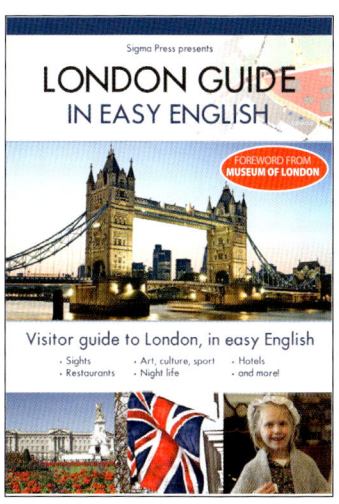

London Guide in Easy English
Patrick Gubbins

The guide covers all the capital's major and minor attractions, hotels, restaurants, parks and green areas and sporting venues, and contains a full directory of necessary information for visitors to London, including advice on working in the city. One of the book's themes is the amazing variety of activities on offer in London, some covered by no other guide, such as whitewater rafting, craft workshops, ski-ing on real snow, visits to courtrooms to watch real trials, and even how to see members of the Royal Family! Packed with exciting ideas and stunning photography, *London Guide in Easy English* is the ideal travel companion for the many visitors to London looking for a guide book written at an easy level of English.
£9.99

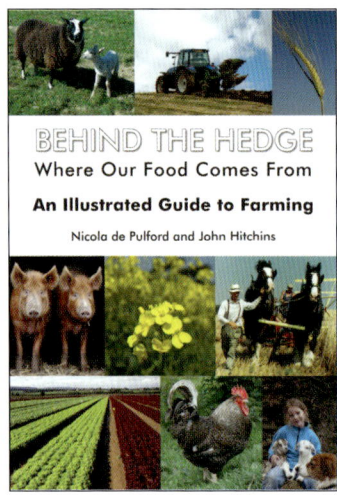

BEHIND THE HEDGE
Where Our Food Comes From
An Illustrated Guide to Farming
Nicola de Pulford & John Hitchins

Behind the Hedge is for everyone who wants to know more about the food we eat, the land it is grown and reared on, and those who farm it. It is an easy-to-follow guide which will help you identify in their natural environment our crops, fruit and farm animals, agricultural buildings and machinery, the farming landscape and the wildlife it supports. By dipping in and out of this beautifully illustrated book, you will learn to recognise the crops, farm animals and wildlife on the other side of the hedge.

£12.99

Walks in the Footsteps of Winston Graham's Poldark
Sue Kittow

Winston Graham was so good at evoking the real landscape of Cornwall, the Cornish people and the unpredictability of the Cornish weather. Feature in page turning plots and empathetic characters, and you have a winner, which the Poldark books have proved to be. Enjoy these walks and learning more about the Poldark places, characters, and history. Each walk includes details of maps, refreshments, history, points of interest and clear directions and sketch map. What makes these walks different is their personal style, the delightful details and excellent photographs, which combine to make this a unique book to keep and pore over.

£8.99

Walks in the Footsteps of Cornish Writers
Sue Kittow

A fascinating insight to contemporary authors and their favourite walks as well as the places that were so special to those well known writers who are no longer with us - and why they were so special. Some writers have made their favourite places famous through verse or novels, others, use regular walks as a valuable part of their writing day. Each walk has a factbox with all essential information, and details of maps, refreshments, history, points of interest and clear directions.
£8.99

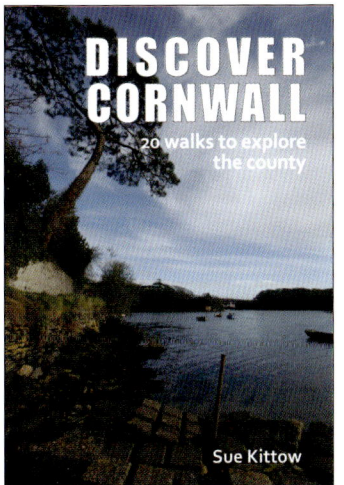

Discover Cornwall
20 walks to explore the county
Sue Kittow

Cornwall's fine golden sands have provided the backdrop for many childhood holidays, but it is also a wonderful county to explore on foot. As well as the coastal footpath, there are numerous less known routes that are great fun to investigate. There are a good range of gentle to moderate walks between 4 and 6 miles in length. Discover Cornwall lists 20 walks providing a healthy and entertaining way to keep fit, learn about Cornwall, and enjoy the beaches, moorland and hisotry of this magical county. The walks have clear directions, delightful details and excellent photographs, maing this a unique book to keep and pore over for readers as well as walkers.
£8.99

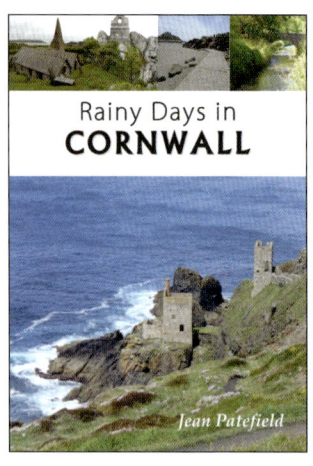

Rainy Days in Cornwall
Jean Patefield

Cornwall has a long and beautiful coastline, with wonderful beaches many of which are excellent for surfing. There are also picturesque valleys and woodland. Overall it merits its place as the premier summer resort in Britain.Unfortunately, being in the west of England, even in high summer wind and rain can lash the beaches, the temperature can plummet and the coast can be shrouded in mist and drizzle. What should one do when your week's summer holiday is turning into a disaster? Carry on regardless, huddled behind a windbreak trying to keep warm or patronise the numerous attractions and spend a fortune? *Rainy Days in Cornwall* offers a solution to this problem with twenty suggestions of free and interesting things to do in Cornwall in less than perfect weather.
£8.99

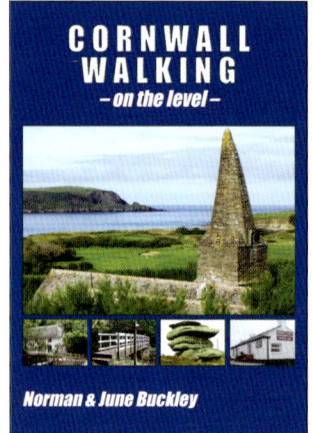

Cornwall Walking on the level
Norman & June Buckley

This book selects and illustrates 28 routes, mainly circular, which explore some of the finest parts of the county, without serious ascent. In addition to the route directions, the distance, ascent, car parking, refreshment and map, with a succinct assessment, are provided for each walk.
£8.99

All of our books are all available on-line at **www.sigmapress.co.uk** or through booksellers. For a free catalogue, please contact:

**Sigma Leisure, Stobart House, Pontyclerc, Penybanc Road, Ammanford, Carmarthenshire SA18 3HP
Tel: 01269 593100   Fax: 01269 596116
info@sigmapress.co.uk   www.sigmapress.co.uk**